CHEMICAL CALCULATIONS

(EIGHTEENTH EDITION)

By

GEORGE I. SACKHEIM
UNIVERSITY OF ILLINOIS AT CHICAGO
(emeritus)

ISBN 1-58874-542-2

Published by
Stipes Publishing L.L.C.

204 W. University Ave.
Champaign, Illinois 61820

PREFACE

This book has been written for the average student in general chemistry. Its purpose is twofold: first, to show how to solve a chemical problem and second, to explain certain concepts, such as atomic structure, oxidation numbers, nomenclature, etc., with which the student has the most trouble in understanding.

The method of logical reasoning is used throughout the book instead of merely substituting in a mathematical formula with the hope that the student will learn to "think through" the problems instead of relying on memorizing certain formulas.

Emphasis is also placed on the use of significant figures.

G.I.S.

18th Edition

January 2006

TABLE OF CONTENTS

TABLE OF CONTENTS *(continued)*

TABLE OF CONTENTS *(continued)*

Chapter 1

GENERAL MATHEMATICAL REVIEW

Percentage

Percent actually means "per hundred" or the parts of one thing in 100 parts of the group. Thus, if we had a mixture consisting of 40 grams of sand and 50 grams of salt, there would be

40 grams of sand in 90 grams of the mixture, and

50 grams of salt per 90 grams of the mixture.

40 grams of sand per 90 grams of the mixture could be written as

$$\frac{40 \text{ grams sand}}{90 \text{ grams mixture}}$$ and dividing both numbers by 90 we have

$$\frac{0.44 \text{ grams sand}}{1 \text{ gram mixture}}.$$ For 100 grams of the mixture there would be

$$100 \times 0.44 \text{ or } 44 \text{ grams sand}, \left(100 \text{ grams mixture} \times \frac{0.44 \text{ grams sand}}{1 \text{ gram mixture}}\right) \text{ or}$$

the mixture would contain 44 percent sand (44 parts per hundred are sand).

Likewise, the percent salt in the mixture $= \dfrac{50 \text{ grams salt}}{90 \text{ grams mixture}} \times 100\% = 56$ percent.

Thus, the percent $= \dfrac{\text{the amount of the substance whose percent you want}}{\text{total amount present}} \times 100\%$ because

$\dfrac{\text{amount you have}}{\text{total amount}}$ = parts per 1 unit and when multiplied by 100 gives parts per 100 which is percent.

Therefore, in any substance, the sum of the percents of all the constituents must always equal 100 percent, or, the sum of all the parts must always equal the whole.

Example No. 1-1. A 20 gram sample of carbon dioxide contains 5.4 grams carbon.

What is the percent carbon in the compound?

The percent carbon $= \dfrac{\text{wt. of carbon}}{\text{total weight}} \times 100\% = \dfrac{5.4 \text{ grams}}{20 \text{ grams}} \times 100\% = 27$ percent.

NOTE: 27 percent, 0.27, $\dfrac{27}{100}$ all mean the same thing.

Example No. 1-2. A certain compound contains 10 grams calcium, 3 grams carbon and 12 grams oxygen. What is the percent calcium in the compound? the percent carbon? the percent oxygen?

Percent calcium $= \dfrac{\text{amount calcium present}}{\text{amount compound present}} \times 100\% = \dfrac{10 \text{ grams}}{(10+3+12) \text{ grams}} \times 100\% = \dfrac{10}{25} \times 100 = \quad 40\%$

Percent carbon $= \dfrac{\text{amount carbon present}}{\text{amount compound present}} \times 100\% = \dfrac{3 \text{ grams}}{(10+3+12) \text{ grams}} \times 100\% = \dfrac{3}{25} \times 100 = \quad 12\%$

Percent oxygen $= \dfrac{\text{amount oxygen present}}{\text{amount compound present}} \times 100\% = \dfrac{12 \text{ grams}}{(10+3+12) \text{ grams}} \times 100\% = \dfrac{12}{25} \times 100 = \quad \underline{48\%}$

$$= 100\%$$

Example No. 1-3   An ore contains 92.5% iron oxide. How much iron oxide is present in 60.0 tons of that ore?

92.5% iron oxide means that there are 92.5 parts of iron oxide for every 100 parts of ore present, or 0.925 parts of iron oxide for each part of ore present.

Then 60.0 tons of ore will contain

$$60.0 \text{ tons of ore} \times \frac{0.925 \text{ parts iron oxide}}{1 \text{ part ore}} = 55.5 \text{ tons of iron oxide present.}$$

Example No. 1-4.   How much iron can be produced from the iron oxide in the previous example if the iron oxide contains 70.0% iron?

From the previous example, we have 55.5 tons of iron oxide. Since this oxide contains only 70.0% iron, the amount of iron present in that iron oxide is

70.0% of 55.5 tons or 0.700 × 55.5 tons or 38.8 tons of iron (to one decimal place).

SIGNIFICANT FIGURES

All measurements are necessarily approximations because no measuring device is absolutely perfect. By convention, a mass of 30.2 grams indicates that the mass was measured to the nearest tenth of a gram and a mass of 30.20 grams indicates an accuracy to the nearest hundredth of a gram. 30.2 contains 3 significant figures (3, 0, and 2) while 30.20 contains 4 significant figures (3, 0, 2, and 0). A significant figure is one that is known to be fairly reliable.

The following rules may be used in determining the number of significant figures:

1. *All digits that are not zero are significant*, regardless of the location of a decimal point.

    3.76 contains 3 significant figures.

    37.6 contains 3 significant figures.

    175.624 contains 6 significant figures.

2. *Zeros between nonzero digits are significant.*

    30.9 contains 3 significant figures.

    4008.27 contains 6 significant figures.

3. *Zeros to the left of the first nonzero digit are not significant.* They indicate the placement of the decimal point.

    0.917 contains 3 significant figures.

    0.0051 contains 2 significant figures.

4. *Zeros to the right of a decimal point are significant if they are at the end of a number.*

    63.0 contains 3 significant figures.

    16.000 contains 5 significant figures.

5. *When a number ends in a zero or zeros that are not to the right of a decimal point, the ending zero or zeros may or may not be significant.*

Thus, 760 may or may not contain 3 significant figures, depending upon the accuracy of the measurement. To avoid any ambiguity in the number of significant figures, such a number should be written using exponential notation.

760 (3 significant figures) is written as 7.60×10^2

760 (2 significant figures) is written as 7.6×10^2

14,700 (3 significant figures) is written as 1.47×10^4.

Rounding Off

To round off a number, we use the following rules:

1. When the number dropped is less than 5, the preceding number remains unchanged, Thus:

    1.6372 is rounded off to 1.637 and

    0.051794 is rounded off to 0.05179, etc.

2. When the number dropped is more than 5, 1 is added to the preceding number. Thus:

    3.7168 is rounded off to 3.717 and

    15.3149 is rounded off to 15.315, etc.

3. When the number dropped is exactly 5, if the preceding number is even, it remains unchanged; if the preceding number is odd, 1 is added to it. Thus:

    7.185 is rounded off to 7.18 and

    0.3735 is rounded off to 0.374, etc.

Thus, rounding off successively, we have

7.1791545, 7.179154, 7.17915, 7.1792, 7.179, 7.18, 7.2, 7.

Addition and Subtraction

In addition and in subtraction, the answer should be rounded off so as to contain the same number of decimal places as in the number with the least number of decimal places.

Thus to add:
$$\begin{array}{r} 17.155 \\ \underline{4.173} \\ 21.328 \end{array}$$ (to be rounded off if necessary)

the least number of decimal places is 3, so the answer should have 3 decimal places, or

21.328 (correct answer)

To add:
$$\begin{array}{r} 12.7 \\ 4.716 \\ \underline{13.85} \\ 31.266 \end{array}$$ (to be rounded off)

the least accurate figure has 1 decimal place so the answer can contain figures only up to the one decimal place and thus the correct result should be

31.3 (correct answer).

Multiplication and Division

In multiplication and in division, the answer should be rounded off so as to contain only as many significant figures as are contained in the least accurate number. Thus:

$$3.17 \times 5.186 = 16.4 \text{ (3 significant figures)}$$

$$183.5 \div 0.717 = 2559 = 2.56 \times 10^3 \text{ (3 significant figures)}$$

Problem Assignment

How many significant figures are contained in each of the following numbers?

1-1.	519	(3)	1-6.	0.0498	
1-2.	0.19	(2)	1-7.	6.35×10^4	
1-3.	659.42	(5)	1-8.	0.10082	
1-4.	2.0050	(5)	1-9.	5 thousand	
1-5.	4×10^{-2}	(1)	1-10.	2.7 millionths	

Perform the following indicated operations, giving an answer with the proper number of significant figures.

1-11. Add: 37.76
             3.907
             226.4   (268.1)

1-12. Subtract: 319.15
                  32.614   (286.54)

1-13. Add: 104.632
           27.09542
           3.6125   (135.340)

1-14. 3.156×4.103 (12.95)

1-15. $12.16 \div 3.12$ (3.90)

1-16. $1.35 \times 2.79 \times 4.06$

1-17. $(1.75)^3$

1-18. 2.176×2.74

1-19. $0.005763 \div 0.0251$

1-20 127.0×0.000352

1-21. Add: 1253
              429.6
             37.254

1-22. Add: 37.0986
           17.93724

1-23. Subtract: 20.6
              3.71

1-24. 0.00518×3.0

DIMENSIONAL ANALYSIS

The dimensions of a number tell in what units it is expressed. Thus, when we say an object has a density of 2 grams per milliliter or 2 g/mL, we are saying that the dimensions are grams per milliliter or g/mL. Dimensions may be added, subtracted, multiplied, and divided just as though they were numbers.

It is very useful in chemical calculations to carry these dimensions throughout the problem because, if done properly, the final answer should have the desired dimensions.

Thus, to find the volume of a 50 gram object of density 2.0 grams per cubic centimeter, we use the relationship

$$\text{Density} = \frac{\text{Mass}}{\text{Volume}} \text{ and so}$$

$$\text{Volume} = \frac{\text{Mass}}{\text{Density}} = \frac{50 \text{ grams}}{2.0 \text{ grams/cubic centimeter}} = 25 \text{ cubic centimeters}$$

where the answer has the dimensions cubic centimeter, which is a unit of volume as was wanted.

Also, if we had made an error in setting up this problem (or any other problem) or if we had solved incorrectly for the volume in the equation, the final answer would have had some dimensions other than of volume and so obviously would be incorrect. Therefore, it is always helpful in solving problems to carry the dimensions along and if, in the final answer, these dimensions are of the right type, then the chances are that the problem was set up properly. (Refer to Chapters 2 and 6 for more examples.)

EXPONENTS

In any chemistry course, very large and very small numbers are expressed exponentially. Actually, the process is very simple and will be quite useful in other courses. Very large or very small numbers are expressed as 10 raised to some power. A power, or exponent, tells how many times a number is repeated as a factor. Thus:

$$10^2 \text{ (ten repeated as a factor 2 times)} = 10 \times 10 = 100$$

$$10^3 \text{ (ten repeated as a factor 3 times)} = 10 \times 10 \times 10 = 1000$$

$$10^5 \text{ (ten repeated as a factor 5 times)} = 10 \times 10 \times 10 \times 10 \times 10 = 100,000$$

By definition, any number raised to the zero power = 1, so $10^0 = 1$. Also, by definition, a number with a negative exponent indicates the reciprocal of the same number with a positive exponent. Thus:

$$10^{-1} = \frac{1}{10^1} = \frac{1}{10} = 0.1 \text{ and}$$

$$10^{-3} = \frac{1}{10^3} = \frac{1}{1000} = 0.001$$

Combining all this in a table we have:

$10^6 =$	1,000,000	$10^{-1} = 0.1$	
$10^5 =$	100,000	$10^{-2} = 0.01$	
$10^4 =$	10,000	$10^{-3} = 0.001$	
$10^3 =$	1000	$10^{-4} = 0.0001$	
$10^2 =$	100	$10^{-5} = 0.00001$	
$10^1 =$	10	$10^{-6} = 0.000001$	
$10^0 =$	1		

Numbers which are not integral (whole number) powers of 10 may be written as a product of two numbers, one of which is a power of 10. The other number is always written with just one figure to the left of the decimal point. Thus:

$$5400 = 5.4 \times 1000 = 5.4 \times 10^3$$

$$627 = 6.27 \times 100 = 6.27 \times 10^2$$

$$0.037 = 3.7 \times 0.01 = 3.7 \times 10^{-2}$$

$$0.00059 = 5.9 \times 0.0001 = 5.9 \times 10^{-4}$$

$$1,637,000 = 1.637 \times 1,000,000 = 1.637 \times 10^6$$

$$0.000000783 = 7.83 \times 0.0000001 = 7.83 \times 10^{-7}$$

Multiplication

When powers of 10 are multiplied, the exponents are added. Thus:

$$10^1 \times 10^3 = 10^{1+3} = 10^4 \text{ since } 10^1 \times 10^3 = 10 \times 1000 = 10,000 = 10^4$$

If each power itself is multiplied by some number, then in multiplication these numbers are multiplied separately and the powers of 10 are added as before. Thus:

$$(4 \times 10^2) \times (2 \times 10^3) = (4 \times 2) \times (10^2 \times 10^3) = 8 \times 10^5$$

$$3.0 \times 10^4 \times 2.3 \times 10^{-1} = (3.0 \times 2.3) \times (10^4 \times 10^{-1}) = 6.9 \times 10^3$$

$$4.2 \times 10^5 \times 6.0 \times 10^7 = (4.2 \times 6.0) \times (10^5 \times 10^7) = 25.2 \times 10^{12} = 2.5 \times 10^{13}$$
$$\text{(2 significant figures)}$$

Division

During division, the powers of 10 are subtracted. Thus:

$$10^4 \text{ divided by } 10^2 = 10^{4-2} = 10^2 \text{ since } \frac{10^4}{10^2} = \frac{10000}{100} = 100 = 10^2$$

If each power is itself multiplied by some number, then in division these numbers are divided separately and the powers of 10 are subtracted as before. Thus:

$$\frac{4.00 \times 10^7}{3.00 \times 10^3} = \frac{4.00}{3.00} \times 10^{7-3} = \frac{4.00}{3.00} \times 10^4 = 1.33 \times 10^4$$

$$\frac{9.0 \times 10^4}{2.0 \times 10^{-6}} = \frac{9.0}{2.0} \times 10^{4-(-6)} = \frac{9.0}{2.0} \times 10^{10} = 4.5 \times 10^{10}$$

Note: $\frac{10^1}{10^1} = 10^{1-1} = 10^0 = 1 \left(\frac{10}{10} = 1 \right)$.

Addition and Subtraction of Exponential Numbers

Exponential numbers may be added or subtracted if the powers of 10 are the same. Thus:

$$4 \times 10^3 + 3 \times 10^3 = (4 + 3) \times 10^3 = 7 \times 10^3$$

$$9 \times 10^{-5} - 5 \times 10^{-5} = (9-5) \times 10^{-5} = 4 \times 10^{-5}$$

$$7 \times 10^{-2} + 4 \times 10^{-2} = (7 + 4) \times 10^{-2} = 11 \times 10^{-2} = 1.1 \times 10^{-1}$$

If the numbers have different powers of 10, then these powers must be equalized before addition or subtraction. Thus:

$\underline{4.0 \times 10^3 + 3 \times 10^2 = ?}$

First let us change the smaller number to one with the same power as the larger.

$$3 \times 10^2 = 0.3 \times 10^3 \text{ then,}$$

$$4.0 \times 10^3 + 0.3 \times 10^3 = 4.3 \times 10^3$$

$\underline{8 \times 10^5 - 6.0 \times 10^6 = ?}$

By changing the smaller exponent to the larger, we have
$$8 \times 10^5 = 0.8 \times 10^6 \text{ and then}$$
$$0.8 \times 10^6 - 6.0 \times 10^6 = -5.2 \times 10^6$$

Raising To a Power

When exponential numbers are to be raised to a power, the number is raised to the proper power and the exponent is multiplied by that power. Thus:

$$(2 \times 10^3)^2 = (2)^2 \times (10^3)^2 = 4 \times 10^{3 \times 2} = 4 \times 10^6$$

$$(4.0 \times 10^4)^3 = (4.0)^3 \times (10^4)^3 = 64 \times 10^{4 \times 3} = 64 \times 10^{12} = 6.4 \times 10^{13}$$

Taking a Root

When taking a root of an exponential number, take the root of the number and divide the power by the number indicated by that root. Thus:

$$\sqrt[2]{4 \times 10^6} = \sqrt[2]{4} \times 10^{6 \div 2} = 2 \times 10^3$$

$$\sqrt[3]{9.00 \times 10^{12}} \times \sqrt[3]{9.00} \times 10^{12 \div 3} = 2.08 \times 10^4$$

Problem Assignment

Express the following as a number multiplied by 10 raised to some power. (Three significant figures.)

1-25.	3186	(3.19×10^3)
1-26.	850,010	(8.50×10^5)
1-27.	0.0917	(9.17×10^{-2})
1-28.	0.000539	(5.39×10^{-4})
1-29.	4,124,000	
1-30.	395,127,689	
1-31.	0.00000826	
1-32.	127.2	
1-33.	68.69	
1-34.	0.0091254	

Change the following exponential numbers to common numbers

1-35.	4×10^3	(4000)
1-36.	6×10^{-2}	(0.06)
1-37.	3.6×10^4	(36,000)
1-38.	5.37×10^{-5}	(0.0000537)
1-39.	4.06×10^{-1}	(0.406)
1-40.	9.27×10^{-2}	
1-41.	3.39×10^6	
1-42.	4.96×10^{-1}	
1-43.	8.07×10^{-8}	
1-44.	3.68×10^0	

Perform the indicated operations.

1-45.	$2 \times 10^3 \times 3 \times 10^4$	(6×10^7)
1-46.	$7.2 \times 10^4 \div 8.6 \times 10^5$	(8.4×10^{-2})
1-47.	$(2.0 \times 10^3)^3$	(8.0×10^9)
1-48.	$3.5 \times 10^3 + 2.5 \times 10^3$	(6.0×10^3)
1-49.	$9.06 \times 10^{-5} \div 3.00 \times 10^{-3}$	(3.02×10^{-2})
1-50.	$\sqrt{6.4 \times 10^9}$	(8.0×10^4)
1-51.	$1.6 \times 10^2 \times 2.8 \times 10^4$	(4.5×10^6)
1-52.	$9.7 \times 10^{-3} \div 1.8 \times 10^4$	(5.4×10^{-7})
1-53.	$6.4 \times 10^4 - 8.2 \times 10^3$	(5.6×10^4)
1-54.	$7.3 \times 10^{-5} - 6.1 \times 10^{-6}$	(6.7×10^{-5})
1-55.	$3.7 \times 10^4 \times 8.2 \times 10^2$	
1-56.	$2.5 \times 10^{-2} \times 2.0 \times 10^{-3}$	
1-57.	$(7.1 \times 10^2)^4$	

1-58.	$\sqrt[3]{2.7 \times 10^{-11}}$
1-59.	$3 \times 10^4 + 2.00 \times 10^6$
1-60.	$2.7 \times 10^{-4} - 5.8 \times 10^{-4}$
1-61.	$3.75 \times 10^6 \times 2.80 \times 10^{-2}$
1-62.	$(1.02 \times 10^3)^4$
1-63.	$3.95 \times 10^4 \div 8.27 \times 10^5$
1-64.	$6.75 \times 10^2 \div 1.86 \times 10^0$
1-65.	$7.20 \times 10^{-2} - 3.75 \times 10^{-2}$
1-66.	$6.02 \times 10^{11} \times 1.09 \times 10^{-6}$
1-67.	$3.57 \times 10^6 \div 4.26 \times 10^4$
1-68.	$\sqrt{3.60 \times 10^7}$
1-69.	$3.19 \times 10^2 + 2.968 \times 10^2$
1-70.	$2.0 \times 10^4 - 2 \times 10^3$

PERCENTAGE ERROR

The percentage error between two results is calculated as follows:

$$\text{percentage error} = \frac{\text{difference between the two results}}{\text{accepted result}} \times 100$$

Example No. 1-5. 10 liters of a gas are compressed to a volume of 2.04 liters. According to Boyle's Law, the volume should be 2.00 liters. Express the error as a percentage deviation from Boyle's Law.

$$\text{percentage error} = \frac{\text{difference}}{\text{accepted result}} \times 100 = \frac{(2.04 - 2.00) \text{ liters}}{2.00 \text{ liters}} \times 100 = \frac{0.04}{2.00} \times 100 = 2\%$$

(Note that the answer contains only 1 significant figure, as required by the difference 0.04.)

Chapter 2

UNITS OF MEASUREMENT

I. S I UNITS

We are all familiar with the common units of measurement such as the inch, the foot, the mile, the ounce, the pound, and many others in the English system of measurement. However, in chemistry we use the Systemé Internationale, the S.I. system, which is a modernized version of the older metric system. Metric units are related decimally; that is, by powers of 10.

A. Base Units

The base SI units are:

base unit	name	abbreviation
Length	Meter	m
Mass	Kilogram	kg
Temperature	Kelvin	K
Time	Second	s
Amount of a substance	Mole	mol
Electrical current	Ampere	A
Luminous intensity	Candela	cd

B.   Common Units

In both the S.I. and metric systems, we use prefixes to designate the various multiples of 10.

The most commonly used prefixes are:

milli	which means 1/1000	or 10^{-3}
centi	which means 1/100	or 10^{-2}
deci	which means 1/10	or 10^{-1}
kilo	which means 1,000	or 10^3
mega	which means 1,000,000	or 10^6

Other prefixes are:

micro	which means 10^{-6}
nano	which means 10^{-9}
pico	which means 10^{-12}

Thus, a millimeter is 1/1000 of a meter; a centimeter is 1/100 of a meter; a decimeter is 1/10 of a meter; a kilometer is 1000 meters, and a nanometer is 10^{-9} meters.

Metric units still in common use are the GRAM, which is about 1/30 of an ounce, and the LITER, which is approximately one quart. Another unit of volume is the cubic centimeter. Thus, a milliliter is 1/1000 of 1 liter, a kilogram is 1000 grams.

It is usually much simpler to abbreviate these units than to write them out, so we use the following abbreviations:

meter*   . . . . m
gram   . . . . . g
liter   L

The prefixes are also abbreviated as follows:

milli . . . . m	centi . . . c	micro . . . μ	pico . . . p
deci . . . . d	kilo . . . . k	nano . . . . n	

Here are few examples of these terms. You should familiarize yourself with them.

meter . . . . m	gram . . . . g	liter L
millimeter . . . . mm	milligram . . . . mg	milliliter . . . . mL
centimeter . . . . cm	centigram . . . . cg	centiliter cL
decimeter . . . . dm	decigram . . . . dg	deciliter . . . . dL
kilometer . . . . km	kilogram . . . . kg	kiloliter . . . . kL
micrometer . . . . . . μm	nanogram . . . . ng	picoliter . . . . pL

*In some textbooks the unit of length is spelled metre.

Comparison of Units

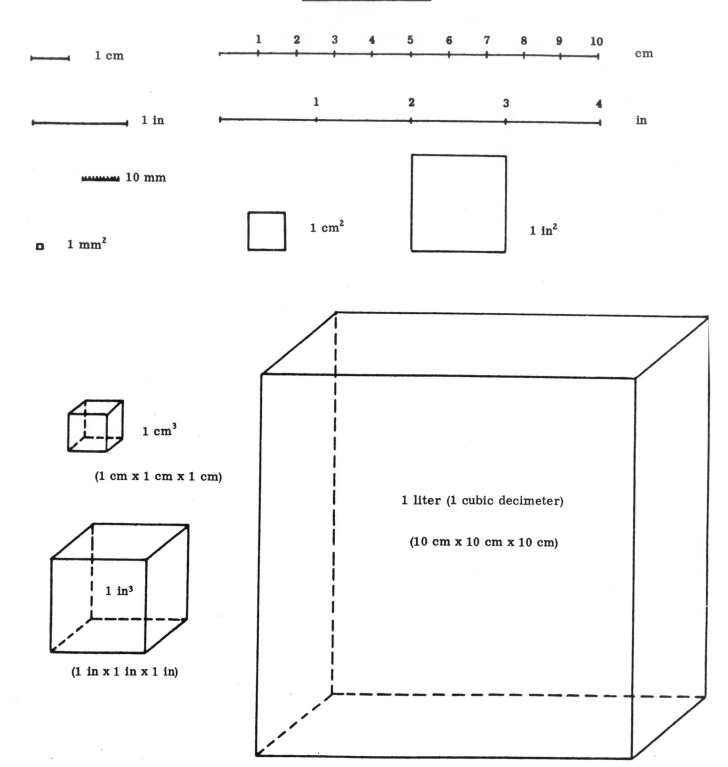

1 cm

1 2 3 4 5 6 7 8 9 10   cm

1 in

1 2 3 4   in

10 mm

1 mm²

1 cm²

1 in²

1 cm³

(1 cm x 1 cm x 1 cm)

1 in³

(1 in x 1 in x 1 in)

1 liter (1 cubic decimeter)

(10 cm x 10 cm x 10 cm)

C. Conversion Units

To change from the English system to the SI or metric system or vice versa, we need some conversion units. There are quite a few of these, but we need only two. They are:

1 inch = 2.54 centimeters or 2.54 cm, and
1 pound = 454 grams or 454 g.

D. Metric-Metric Conversions

Example No. 2-1. 2.50 mL = _____ L.

We know that there are 1000 mL in a liter and we have to decide whether to multiply or divide by this number. However, if we are careful with the units, there should be no question as to which is correct. Thus:

$$2.50 \text{ mL} = 2.50 \text{ mL} \times \frac{1 \text{ L}}{1000 \text{ mL}} = 0.00250 \text{ L}$$

where we see that the unwanted units cancel, leaving only the desired one.

Example No. 2-2. 25.4 m = _____ cm.

Since there are 100 cm in a m,

$$25.4 \text{ m} = 25.4 \text{ m} \times \frac{100 \text{ cm}}{1 \text{ m}} = 2540 \text{ cm} = 2.54 \times 10^3 \text{ cm}.$$

Example No. 2-3 16.25 kg = _____ mg.

Here we are converting kilograms to milligrams, but since we have no overall conversion factor, we will have to convert it stepwise. We know that there are 1000 grams in a kilogram and also that there are 1000 milligrams in a gram and if we set up the units properly, they will cancel, leaving the answer in milligrams. Thus:

$$16.25 \text{ kg} = 16.25 \text{ kg} \times \frac{1000 \text{ g}}{1 \text{ kg}} \times \frac{1000 \text{ mg}}{1 \text{ g}} = 16,250,000 \text{ mg} = 1.625 \times 10^7 \text{ mg}.$$

Example No. 2-4. 0.096 cm = _____ mm = _____ m.

We know that there are 10 mm in a cm, since a centimeter is 1/100 of a meter and a millimeter is 1/1000 of a meter. Then, setting it up so that the units cancel,

$$0.096 \text{ cm} = 0.096 \text{ cm} \times \frac{10 \text{ mm}}{1 \text{ cm}} = 0.96 \text{ mm}.$$

When changing millimeters to meters, we know that there are 1000 mm in a m, so

$$0.96 \text{ mm} = 0.96 \text{ mm} \times \frac{1 \text{ m}}{1000 \text{ mm}} = 0.00096 \text{ m}.$$

if we had changed the 0.096 cm to m, then, since there are 100 cm in a meter,

$$0.096 \text{ cm} = 0.096 \text{ cm} \times \frac{1 \text{ m}}{100 \text{ cm}} = 0.00096 \text{ m} \text{ (which is the same as above).}$$

E. English-Metric Conversions

Next we come to the methods of converting from the English system to the metric and vice versa.

Example No. 2-5. If an object weighs 25.0 lbs., how many grams does it weigh?

To solve this problem, we must use a conversion unit for weights. From our list, we see that 1 lb = 454 g. Therefore, when changing 25.0 lbs to g, we must multiply by 454 g/lb., because we can see that there are many more g than lbs. Thus:

$$25.0 \text{ lb} = 25.0 \text{ lb} \times \frac{454 \text{ g}}{1 \text{ lb}} = 11,350 \text{ g} = 1.14 \times 10^4 \text{ g to 3 significant figures}$$

Example No. 2-6. 5.00 yds = _____ m.

For this problem, we need a conversion factor for length, so we use 2.54 cm = 1 in. Therefore,

$$5 \text{ yds.} = 5 \text{ yds.} \times \frac{3 \text{ ft.}}{1 \text{ yd.}} = 15 \text{ ft.} = 15 \text{ ft.} \times \frac{12 \text{ in.}}{1 \text{ ft.}} = 180 \text{ in.}$$ We are writing it in this form to show that the

units must cancel in order to get the proper result. Since there are more cm than in., we multiply by 2.54 to get the number of cm so that

$$5.00 \text{ yds.} = 180 \text{ in.} = 180 \text{ in.} \times \frac{2.54 \text{ cm}}{1 \text{ in.}} = 457 \text{ cm (3 significant figures)}$$

But the problem asks for m, so we must convert cm to m by dividing by 100 since there are 100 cm to a m. Therefore,

$$5.00 \text{ yds.} = 457 \text{ cm} = 457 \text{ cm} \times \frac{1 \text{ m}}{100 \text{ cm}} = 4.57 \text{ m.}$$

Example No. 2-7. 1.00 ton = _____ kg.

In this problem we must convert tons to lbs to g to kg since we do not have one overall conversion factor for changing tons to kg. We know that there are 2000 lb. to a ton and 1000 g to a kg. Since there are more lbs. than tons, we multiply by $\frac{2000 \text{ lb.}}{1 \text{ ton}}$; since there are more g than lb., we multiply by $\frac{454 \text{ g}}{1 \text{ lb.}}$; and since there are fewer kg than g, we multiply by $\frac{1.00 \text{ kg}}{1000 \text{ g}}$. So we have

$$1.00 \text{ ton} = 1.00 \text{ ton} \times \frac{2000 \text{ lb.}}{1 \text{ ton}} \times \frac{454 \text{ g}}{1 \text{ lb.}} \times \frac{1.00 \text{ kg}}{1000 \text{ g}} = 908 \text{ kg.}$$

By carefully cancelling our units, we see that the answer comes out in kg as desired.

Example No. 2-8. What is the volume in liters of a rectangular box 1.0 m long, 50 cm wide, and 20 mm high?

We know that the volume of a box equals length × width × height. However, we can never multiply to get volume unless all of the dimensions have the same units, so let us convert all of these measurements to cm. 1 m = 100 cm, 50 cm = 50 cm, 20 mm = 2.0 cm.

Therefore, when we multiply we get

$$100 \text{ cm} \times 50 \text{ cm} \times 2.0 \text{ cm} = 10,000 \text{ cm}^3$$

Since the problem asks for liters and since there are 1000 cm³ in a liter, the volume of the box is

$$10,000 \text{ cm}^3 \times \frac{1 \text{ L}}{1000 \text{ cm}^3} = 10 \text{ L}$$

Example No. 2-9.   What is the volume in liters of a rectangular box whose dimensions are 2.00 yds. × 40.0 in. × 0.800 m?

Here again the best method is to convert all the units to cm so we can find the volume in cm³ and then change it very easily to liters. Therefore:

$$2.00 \text{ yds.} = 2.00 \text{ yds} \times \frac{36.0 \text{ in.}}{1 \text{ yd.}} = 72.0 \text{ in.} \times \frac{2.54 \text{ cm}}{1 \text{ in.}} = 183 \text{ cm}$$

$$40.0 \text{ in.} = 40.0 \text{ in.} \times \frac{2.54 \text{ cm}}{1 \text{ in.}} = 102 \text{ cm}$$

$$0.800 \text{ m} = 0.800 \text{ m} \times \frac{100 \text{ cm}}{1 \text{ m}} = 80.0 \text{ cm}$$

so the volume will be 183 cm × 102 cm × 80.0 cm = 1,490,000 cm³

$$1,490,000 \text{ cm}^3 \times \frac{1 \text{ L}}{1000 \text{ cm}^3} = 1.49 \times 10^3 \text{ L (3 significant figures)}$$

II. MEASUREMENT OF TEMPERATURE

A. Comparison of the Temperature Scales

While we are all familiar with the Fahrenheit temperature scale, it is seldom used in the field of chemistry. Instead we use the Celsius (Centigrade) and the Kelvin (Absolute) systems. Let us compare these systems and see how they differ. The freezing point of water on the Fahrenheit or F scale is 32°, on the Celsius or C Scale 0°, and on the Kelvin or K scale 273. The boiling point of water on the F scale is 212°, on the C scale 100°, and on the K scale 373. Note that absolute temperatures are given in kelvins and not in degrees.

---212°	------100°	------373	---- Boiling point of water
--- 32°	------ 0°	------273	----Freezing point of water
Fahrenheit F	Celsius C	Kelvin K	

We see that on the F scale, there are 180 units between the freezing and boiling point of water; on the C scale and on the K scale there is a difference of 100 units. It is very important to be able to convert from one system to another, because quite often we take our measurements in one system and do our calculations in another.

B. Fahrenheit-Celsius Conversions

To convert from F to C or vice versa, we use the equations

$$°F = \left(°C \times \frac{1.8°F}{1.0°C}\right) + 32°F \text{ and } °C = \frac{1.0°C}{1.8°F}(°F - 32°F).$$

Let us study a few examples and see how this equation can be used.

Example No. 2-10.   50°C = _____ °F.

$$°F = \left(50°C \times \frac{1.8°F}{1.0°C}\right) + 32°F = 90°F + 32°F = 122°F$$

Example No. 2-11. 180° C = _____ °F.

$$°F = \left(180°C \times \frac{1.8°F}{1.0°C}\right) + 32°F = 324°F + 32°F = 356°F$$

Example No. 2-12. −60°C = _____ °F.

$$°F = \left(-60°C \times \frac{1.8°F}{1.0°C}\right) + 32°F = -108°F + 32°F = -76°F.$$

Here we must be careful to watch the signs, especially since we have negative temperatures.

Example No. 2-13. 140°F = _____ °C.

$$°C = \frac{1.0°C}{1.8°F}(140°F - 32°F)$$

$$°C = \frac{1.0°C}{1.8°F}(108°F) = 60°C.$$

Example No. 2-14. Body temperature is 98.6°F. What is this C?

$$°C = \frac{1.0°C}{1.8°F}(98.6°F - 32°F)$$

$$°C = \frac{1.0°C}{1.8°F}(66.6°F) = 37°C$$

Example No. 2-15. −112°F = _____ °C.

$$°C = \frac{1.0°C}{1.8°F}(-112°F - 32°F)$$

$$°C = \frac{1.0°C}{1.8°F}(-144°F) = -80°C$$

C. Celsius-Kelvin Conversions

To convert from C to K or vice versa, we use the equations

$$K = (°C + 273°C)\frac{1.0\ K}{1.0°C} \text{ and } °C = (K - 273)\frac{1.0°C}{1.0\ K}$$

Example No. 2-16. 28°C = _____ K.

$$K = (28°C + 273°C)\frac{1.0\ K}{1.0°C} = 301\ K.$$

Example No. 2-17. 621 K = _____ °C.

$$°C = (621\ K - 273\ K)\frac{1.0°C}{1.0\ K} = 348°C.$$

Example No. 2-18. 50°F = _____ K.

This is actually two problems because first we must convert from F to C and then from C to K. To find C

$$°C = \frac{1.0°C}{1.8°F}(50°F - 32°F)$$

$$°C = \frac{1.0°C}{1.8°F}(18°F) = 10°C$$

and then

$$K = (10°C + 273°C)\frac{1.0 \text{ K}}{1.0°C} = 283 \text{ K}$$

Example No. 2-19. 135 K = _____ °F.

Here again we have two problems, first in converting K to C and then C to F

$$°C = (135 \text{ K} - 273 \text{ K})\frac{1.0°C}{1.0 \text{ K}} = -138°C$$

$$°F = -138°C \times \frac{1.8°F}{1.0°C} + 32°F = -216.4°F.$$

III. DENSITY

A. Definition

Density is defined as mass per unit volume or algebraically:

$$D = \frac{M}{V}$$

B. Units

There are many possible units for density such as lbs./ft.^{3}, g/mL, kg/L, etc., but the one most commonly used in chemistry is g/mL. We must never use the term density without giving its units because of this possibility of confusion with other systems.

C. Examples of Problems

Example No. 2-20. An object has a mass of 100 g and a volume of 50 cm^3. What is its density?

$$D = M/V = 100 \text{ g}/50 \text{ cm}^3 = 2.0 \text{ g/cm}^3$$

We must be careful to keep the units in the problem as well as in the answer so that there will be no chance for error.

Example No. 2-21. The density of a certain object is 4.00 g/mL and it weighs 1.00 kg. What is its volume?

$$D = M/V \text{ and, therefore, } V = M/D$$

However, the density is in g/mL and the mass in kg so we change the kg to g by multiplying 1.00 kg by $\frac{1000 \text{ g}}{1 \text{ kg}}$ $\left(1.00 \text{ kg} \times \frac{1000 \text{ g}}{1 \text{ kg}} = 1000 \text{ g}\right)$.

$$\text{Then, } V = \frac{M}{D} = \frac{1000 \text{ g}}{4.00 \text{ g/mL}} = 250 \text{ mL}$$

We see here that by carrying the units all the way through the problem and by cancelling in the proper places, we arrive at the unit mL which is a unit of volume as we wanted. If we were looking for volume and the answer came out in g or mL/g or some other units we would immediately know something was wrong because we must come out in the units of the thing we are looking for.

Example No. 2-22. A cubical box is 5 cm on a side and weighs 500 g. What is its density?

$$\text{Volume} = l \times w \times h = 5 \text{ cm} \times 5 \text{ cm} \times 5 \text{ cm} = 125 \text{ cm}^3$$

$$D = M/V = 500 \text{ g}/125 \text{ cm}^3 = 4 \text{ g/cm}^3$$

Example No. 2-23. A container full of mercury weighs 1000 g. The empty container weighs 320 g. The density of the mercury is 13.6 g/cm³. What is the volume of the box?

Mass of mercury plus container	1000 g	$D = M/V$
Mass of container	320 g	$13.6 \text{ g/cm}^3 = \dfrac{680 \text{ g}}{V}$
Mass of mercury	680 g	$V = \dfrac{680 \text{ g}}{13.6 \text{ g/cm}^3} = 50.0 \text{ cm}^3 = \text{volume of mercury} = \text{volume of box}$

Example No. 2-24. A steel ball has a density of 7.2 g/mL and a volume of 5.0 L. What does it weigh?

First, 5.0 L = 5000 mL (since the density is in g/mL, the volume must also be in mL).

$$D = M/V \qquad 7.2 \text{ g/mL} = \frac{M}{5000 \text{ mL}}$$

$$M = 7.2 \text{ g/mL} \times 5000 \text{ mL} = 36,000 \text{ g} = 3.6 \times 10^4 \text{ g}.$$

Example No. 2-25. The density of water is 1.00 g/cm³. What volume will 500 g of it occupy?

Since $D = M/V$

$$1.00 \text{ g/cm}^3 = \frac{500 \text{ g}}{V} \text{ and } V = \frac{500 \text{ g}}{1.00 \text{ g/cm}^3} = 500 \text{ cm}^3$$

Here we see again the simplicity of the metric system—500 g = 500 cm³ (water).

A Comparison of the Weights of One Cubic Centimeter of Various Substances

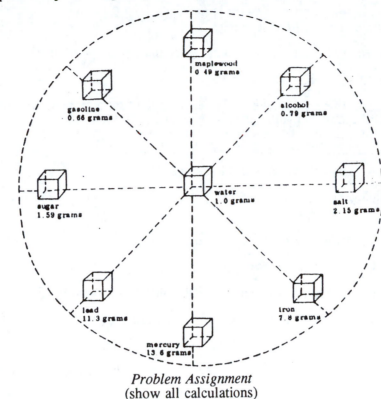

Problem Assignment
(show all calculations)

2-1.   13.85 m = _____ cm = _____ mm = _____ μm

2-2.   31.8 kg = _____ lb.

70.0 lb.

2-3.   40°C = _____ °F

104°F

2-4.   An object has a mass of 155 g and a volume of 27.5 mL. What is its density?

5.64 g/mL

2-5.   339 mL = _____ L = _____ cm³

0.339 L, 339 cm³

2-6.   3.058 g = _____ mg = _____ kg

3.058 × 10³ mg
3.058 × 10⁻³ kg

3.058 g = _____ ng

3.058 × 10⁹ ng

2-7.   821°F = _____ K

711 K

2-8.   600 K = _____ °C = _____ °F

327°C, 621°F

2-9.   What volume of mercury (density 13.6 g/cm³) will have a mass of 0.0375 kg?

2.76 cm³

2-10. The density of bromine is 3.10 g/mL. What will be the mass of 250 mL of it?

775 g

2-11. A piece of iron has a volume of 150.0 mL and has a mass of 1.17 kg. What is its density?

7.80 g/mL

2-12.   159 miles = _____ km   256 km

2-13.   The density of osmium is 22.50 g/cm^3. How many
          pounds does a cubic yard of it weigh? 3.79×10^4 lb

2-14.   The density of osmium is 22.5 g/cm^3. What is the mass of 1.00 m^3 of it?

2-15.   Which is the coldest, 12 K, −450°F or −260°C?

2-16.   The postal rate for a first class letter is 37 cents for the first ounce and 23 cents for each succeeding
          ounce. What will it cost to mail a letter weighing 45 grams?

2-17.   14.7 mm = _____ cm = _____ m = _____km.

2-18.   Aluminum melts at 933 K. What is this in degrees F?

2-19.   A rectangular block 60.0 cm long × 50.0 cm wide × 20.0 cm deep. If its density is 2.70 g/cm^3, what
          is its mass?

2-20.   How many cubic feet are there in a cubic yard?

2-21.   How many cubic inches are there in a cubic foot?

2-22.   How many cubic centimeters are there in a cubic inch? In a cubic foot?

2-23.   A rectangular block is 1.75 cm long, 60.0 cm wide, and 25.0 mm deep. What is its volume in liters?

2-24.   Mercury has a density of 13.6 g/mL. How many lb./ft.^{3} is this?

2-25.   Concentrated sulfuric acid has a density of 1.86 g/cm^3. What volume will 300 g of it occupy?

2-26.   Liquid oxygen boils at −183°C. What is this in °F?

2-27.   The distance between two points is 40.0 miles. How many km is this?

2-28.   A piece of lead (density 11.3 g/mL) has a volume of 450 mL. What is its mass?

2-29.   What is the mass of a cubic meter of silver? (Density = 10.5 g/mL.)

2-30.   A jet plane travels at 2000 mi./hr. How many ft./sec. is this? How many cm/sec.? How many m/sec.?

2-31.   At what temperature will the F and C readings be numerically the same but opposite in sign?

2-32.   At what temperature will the F reading be twice the C reading?

2-33.   At what temperature will the C reading be twice the F reading?

2-34.   Which is heavier, a cubic yard of iridium, density 22.4 g/cm^3 or a cubic meter of platinum, density
          21.4 g/cm^3?

2-35.   A man signs his name on a piece of paper. The signature is found to weigh 3.9 mg. How many pounds
          does it weigh?

2-36. 1.6×10^2 mg = _____ g = _____kg.

2-37. 8.62×10^{-2} g/mL = _____g/L = _____kg/L.

2-38.   125 lb/ft^3 = _____ g/mL.

2-39.   0.154 g/mL = _____ mg/mL = _____mg/L = _____ kg/L.

2-40.   How many cm^3 are there in a rectangular box 50.0 in. long, 40.0 cm wide, and 25.0 mm deep?

2-41.   2.50 μg = _____ mg

2-42.   −120°C = _____ °F = _____ K

2-43.   3.78 cm^3 = _____ L = _____ m^3

2-44.   Which is colder, 100 K or −150°C?

2-45. An individual has a temperature of 99.2°F, a height of 5'4" and a weight of 120 lb. Express these measurements in °C, cm, and kg respectively.

2-46. A patient weighing 150 lb. was told to take 2.0 mg of a medication per kg of body weight. How much medication should be taken?

2-47. 25.6 mg/dL = _____ g/L

2-48. Two cities are 50 km apart. How many miles is this?

2-49. A rectangular block is 3.00 ft. long, 18.0 in. wide and 200 cm high. What is its volume in liters?

2-50. An analytical balance will weigh accurately to 0.001 mg. How many lb. is this?

2-51. How many cm^3 are there in a sphere 4.00 cm in diameter?

2-52. The density of mercury is 13.6 g/mL. Express this in kg/m^3.

2-53. Many medications are measured in mg. If a tablet weighs 1.8×10^{-4} ounces, how many mg does it weigh?

2-54. When a steel sphere weighing 23.4 g is placed in a graduated cylinder containing 20.0 mL of water, the water level rises to 23.1 mL. What is the density of the sphere? What is its radius?

2-55. 18.0 mL of water contains 6.02×10^{23} molecules. What is the volume occupied by one molecule? Assuming the molecules to be spherical, what is the radius of a water molecule?

2-56. If there are 5.00×10^6 red blood cells in 1.00 mm^3 of blood, how many are there in 1.00 mL?

2-57. Light travels at the rate of 1.86×10^6 mi./sec. Express this in km/sec.

2-58. A light year is the distance light travels in one year. Express this distance in miles; in kilometers.

2-59. How many molecules can be placed on the head of a pin whose diameter is 1.50 mm in order to form a layer one molecule thick? Assume the cross-sectional area of a molecule to be 1.50×10^{-19} m.

2-60. The diameter of an atom is about 10^{-8} cm. How many atoms would have to be piled on top of each other in order to make a pile 1.00 in. high?

2-61. 1.59 pg = _____ g

2-62. 0.503 nm = _____ pm

2-63. 3.25 mL = _____ pL

2-64. 1.27×10^{-8} m = _____ pm = _____ nm

2-65. A tiny piece of an alloy has mass of 2.90 ng and a volume of 3.33 pL. What is its density in g/mL?

Chapter 3

NAMING OF CHEMICAL COMPOUNDS

Before we can begin to make progress in chemistry, we must master its vocabulary. While there are many thousands of different chemical compounds, there is a very definite system of nomenclature whereby we can name most compounds simply and easily.

We can divide the compounds into two main types—those which are true binary compounds (that is, contain only two types of elements) and those which contain more than two types of elements.

I. NAMING OF BINARY COMPOUNDS

The names of all compounds containing only two elements end in IDE. Binary compounds may be subdivided into two types—those whose first element is a metal, and those whose first element is a non-metal. In both cases the second element is a non-metal.

For binary compounds whose first element is a metal, we use the following system:

$$\text{name of metal} \quad \text{stem} + \text{IDE}$$

where the stem is merely an abbreviation for the name of the second element (the non-metal). The stems of the most commonly used elements are indicated in the following table:

Stems of the most commonly used elements:

Oxygen	ox	Nitrogen	nitr
Chlorine	chlor	Phosphorus	phosph
Carbon	carb	Fluorine	fluor
Iodine	iod	Bromine	brom
Hydrogen	hydr	Sulfur	sulf

Example No. 3-1. ... Name the compound NaCl.

The symbol Na represents the metallic element sodium. The symbol Cl represents the non-metallic element chlorine, whose stem is chlor. Therefore, the name of NaCl is sodium chlor+IDE or *sodium chloride*. Likewise, for the following compounds.

CaO   is calcium oxide

CaC_2 is calcium carbide

AlN   is aluminum nitride

K_2S is potassium sulfide

HCl   is hydrogen chloride

Note that hydrogen is considered as a metal when it is written first in a binary compound.

There are a few special cases in this system of naming. The ammonium ion NH_4^+ is considered as a simple metallic ion, so

NH_4Cl is ammonium chloride

$(NH_4)_2S$ is ammonium sulfide

There are also two negative ions which are considered to be special cases: the hydroxide (OH^-) and the cyanide (CN^-) are considered as simple negative ions. Thus:

KCN   is potassium cyanide $Mg(OH)_2$ is magnesium hydroxide

NH_4CN is ammonium cyanide

If a metal forms two different ions, the charges on those ions is indicated by Roman numerals. Or, according to the classical system, the ion with the smaller charge is given the ending OUS and the ion with the greater charge is given the ending IC. Thus:

Fe^{2+} is iron (II)   or the ferrous ion

Fe^{3+} is iron (III)   or the ferric ion

Cu^+ is copper (I)   or the cuprous ion

Cu^{2+} is copper II   or the cupric ion

Hg_2^{2+} is mercury (I)   or the mercurous ion

Hg^{2+} is mercury (II)   or the mercuric ion

Sn^{2+} is tin (II)   or the stannous ion

Sn^{4+} is the tin (IV)   or the stannic ion

Other metals that form two different ions are lead and manganese. The rules given on the preceding page apply to compounds of these elements in the same manner.

$FeCl_2$ is iron (II) chloride   or ferrous chloride

CuS   is copper (II) sulfide   or cupric sulfide

Cu_2O is copper (I) oxide   or cuprous oxide

HgO   is mercury (II) oxide   or mercuric oxide

SnO_2 is tin (IV) oxide   or stannic oxide

For binary compounds whose first element is a non-metal, we use the following system:

prefix + name of first element   prefix + stem + IDE

where the stem is again the abbreviation for the second element. In addition, prefixes are used. The prefixes designate how many atoms of the first and second element are present. The prefixes used are as follows:

(mono = 1)   tetra = 4
di   = 2   penta = 5
tri   = 3   hexa   = 6
                         hepta = 7

The prefix "mono" is understood and not written. An exception to this rule is CO, carbon monoxide.

Example No. 3-2.   Name the compound CO_2.

The symbol C represents the element carbon. There is only one carbon atom present, so the prefix "mono" is understood and not written. The symbol O represents the element oxygen whose stem is ox. There are 2 atoms of the second element present so we use the prefix "di." The ending, as with all binary compounds is "ide." Combining all this information, we have:

prefix + name of first element	prefix + stem + ide
carbon	di + ox + ide or *carbon dioxide*.
Likewise, P_2O_3	is diphosphorus trioxide
P_2O_5	is diphosphorus pentoxide
SO_2	is sulfur dioxide
SO_3	is sulfur trioxide
N_2O	is dinitrogen oxide
NO_2	is nitrogen dioxide
CCl_4	is carbon tetrachloride
Cl_2O_7	is dichlorine heptoxide

II. NAMING OF ACIDS DERIVED FROM BINARY COMPOUNDS

RULE: For binary compounds with Hydrogen as the first element, place the term *hydro* at the front of the stem of the second element, the letters IC at the end of the stem and add the word ACID. Thus:

Hbr	(hydrogen bromide)	is hydrobromic acid
H_2S	(hydrogen sulfide)	is hydrosulfuric acid
HF	(hydrogen fluoride)	is hydrofluoric acid

III. NAMING OF ACIDS WHOSE NEGATIVE GROUP CONTAINS OXYGEN

Ternary (oxy) acids contain the elements hydrogen, a non-metal, and oxygen in that order. The most common ternary acid is named as follows: stem of non-metal + IC followed by the word ACID. Thus, the most common acid of sulfur is sulfuric acid; of nitrogen, nitric acid. The student must remember which is the most common acid derived from several elements. The only ones which we are concerned with are the common acids of phosphorus, sulfur, nitrogen, chlorine, and carbon.

H_2SO_4	H_3PO_4	HNO_3	$HClO_3$	H_2CO_3
sulfur*ic acid*	phosphor*ic acid*	nitr*ic acid*	chlor*ic acid*	carbon*ic acid*

If the acid contains ONE LESS OXYGEN ATOM than the most common one, the ending on the stem is changed from IC to *OUS*.

H_2SO_3	H_3PO_3	HNO_2	$HClO_2$
sulfur*ous acid*	phosphor*ous acid*	nitr*ous acid*	chlor*ous acid*

If the acid contains TWO LESS OXYGEN ATOMS than the most common one, the ending on the stem is changed from IC to OUS and the prefix *HYPO* is added.

H_3PO_2	$(HNO)_2$	HClO
*hypo*phosphor*ous acid*	*hypo*nitr*ous acid*	*hypo*chlor*ous acid*

If the acid contains ONE MORE OXYGEN ATOM than the most common one, the prefix *PER* is added to the name of the most common acid.

Thus, $HClO_4$ is *per*chloric acid

Combining all this in a table, we have:

			$HClO_4$ perchloric acid
H_2SO_4 sulfuric acid	H_3PO_4 phosphoric acid	HNO_3 nitric acid	$HClO_3$ chloric acid
H_2SO_3 sulfurous acid	H_3PO_3 phosphorous acid	HNO_2 nitrous acid	$HClO_2$ chlorous acid
	H_3PO_2 hypophosphorous acid	$(HNO)_2$ hyponitrous acid	$HClO$ hypochlorous acid

IV. NAMING OF SALTS CORRESPONDING TO THE OXY-ACIDS

For salts derived from the most common acids (the IC acids) remove the ending IC from the stem, add the letters *ATE* and prefix the name of the positive ion. Thus,

Na_2SO_4 (derived from H_2SO_4—sulfuric acid) is sodium sulf*ate*.

KNO_3 (derived from HNO_3—nitric acid) is potassium nitr*ate*.

$(NH_4)_3PO_4$ (derived from H_3PO_4—phosphoric acid) is ammonium phosph*ate*.

$Ca(ClO_3)_2$ (derived from $HClO_3$—chloric acid) is calcium chlor*ate*.

For salts derived from the OUS acids, drop the ous ending from the stem, add the letters *ITE* and prefix the name of the positive ion. Thus,

K_2SO_3 (derived from H_2SO_3—sulfurous acid) is potassium sulf*ite*.

$Al(NO_2)_3$ (derived from HNO_2—nitrous acid) is aluminum nitr*ite*.

$KClO_2$ (derived from $HClO_2$—chlorous acid) is potassium chlor*ite*.

For salts derived from the hypo----ous acids, drop the ous ending from the stem, add the letters *ITE* and prefix the name of the positive ion (keeping the term hypo).

$KClO$ (derived from $HClO$—hypochlorous acid) is potassium hypochlor*ite*.

For salts derived from the per---ic acids, drop the IC ending from the stem, add the letters *ATE* (keeping the per) and prefix the name of the positive ion.

$KClO_4$ (derived from $HClO_4$—perchloric acid) is potassium perchlor*ate*.

			$Mg(ClO_4)_2$ magnesium perchlorate
Na_2SO_4 sodium sulfate	K_3PO_4* potassium phosphate	$Ca(NO_3)_2$ calcium nitrate	$Mg(ClO_3)_2$ magnesium chlorate
Na_2SO_3 sodium sulfite		$Ca(NO_2)_2$ calcium nitrite	$Mg(ClO_2)_2$ magnesium chlorite
			$Mg(ClO)_2$ magnesium hypochlorite

* For special rules on the naming of phosphorus compounds, consult your text.

V. NAMING OF SALTS CONTAINING MORE THAN ONE POSITIVE ION

a) Salts containing 2 positive ions, one of which is hydrogen. Construct the name as follows:

1. Give the name of the positive ion other than hydrogen.
2. The letters *bi* may be used to indicate the hydrogen ion.
3. Give the proper name for the negative ion using the above rules.

Thus:

$NaHSO_4$	is sodium hydrogen sulfate, or sodium bisulfate
$KHSO_4$	is potassium hydrogen sulfate, or potassium bisulfate
$LiHSO_3$	is lithium hydrogen sulfite, or lithium bisulfite
$NaHCO_3$	is sodium hydrogen carbonate, or sodium bicarbonate
$Mg(HCO_3)_2$	is magnesium hydrogen carbonate, or magnesium bicarbonate

b) Phosphate salts containing more than 1 type of positive ion, one of which is hydrogen.

1. Give the name of the first positive ion using prefixes to indicate how many atoms of it are present.
2. Give the name of second positive ion using prefixes to indicate how many atoms of it are present.
3. Give the proper name for the negative ion—phosphate in this case.

NaH_2PO_4	is sodium dihydrogen phosphate
K_2HPO_4	is dipotassium hydrogen phosphate

Note that the prefix "mono" is usually understood and not written.

Problem Assignment

Name the following compounds:

3-1. K_2S	3-18. $Al_2(SO_4)_3$	3-35. $MgSO_4$
3-2. $MgBr_2$	3-19. $SnCl_4$	3-36. $Ca(BrO_2)_2$
3-3. $Ca(ClO_3)_2$	3-20. $AsCl_3$	3-37. SiO_2
3-4. $CaSO_4$	3-21. ICl	3-38. $CuCl$
3-5. $AgCl$	3-22. NH_4OH	3-39. $KClO_2$
3-6. BaS	3-23. $Fe(ClO_4)_2$	3-40. Na_2SO_3
3-7. Mg_3P_2	3-24. HNO_2	3-41. KF
3-8. H_2CO_3	3-25. CS_2	3-42. P_2O_3
3-9. $Mg_3(PO_4)_2$	3-26. $CuCl_2$	3-43. $HClO$
3-10. $LiCl$	3-27. PCl_3	3-44. NO_2
3-11. MgI_2	3-28. $Ca(NO_3)_2$	3-45. NaH
3-12. $Al(NO_2)_3$	3-29. KH_2PO_4	3-46. P_2S_3
3-13. $Zn(OH)_2$	3-30. $Cu(CN)_2$	3-47. $Pb(NO_3)_2$
3-14. SnI_2	3-31. $KHCO_3$	3-48. H_2Se
3-15. $AsCl_5$	3-32. $NaHSO_4$	3-49. H_3PO_2
3-16. $CuSO_3$	3-33. Li_2HPO_4	3-50. CaH_2
3-17. HF	3-34. H_3PO_3	

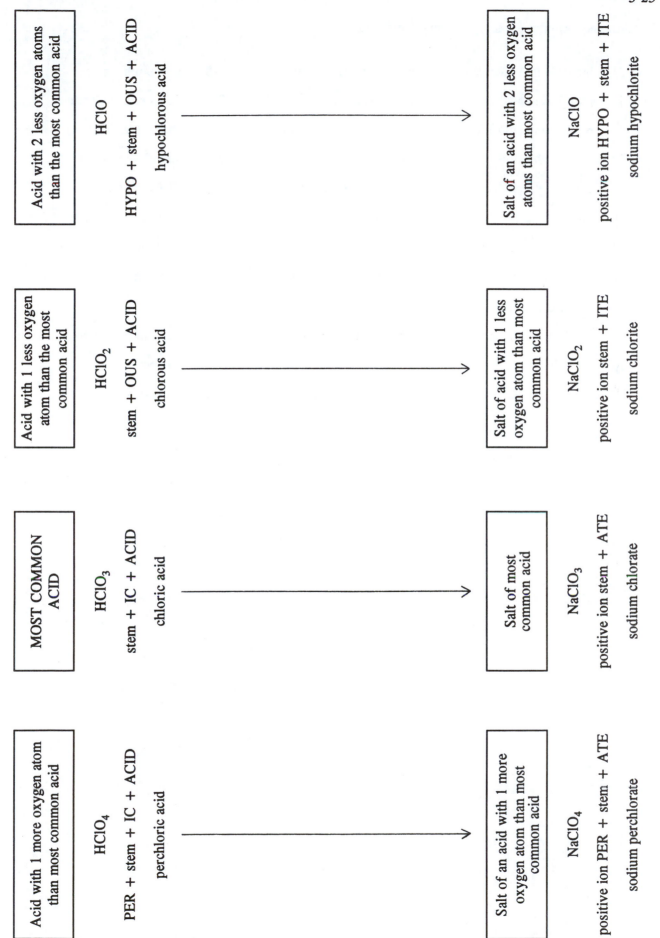

Acid with 1 more oxygen atom than most common acid

HClO₄

PER + stem + IC + ACID

perchloric acid

MOST COMMON ACID

HClO₃

stem + IC + ACID

chloric acid

Acid with 1 less oxygen atom than the most common acid

HClO₂

stem + OUS + ACID

chlorous acid

Acid with 2 less oxygen atoms than the most common acid

HClO

HYPO + stem + OUS + ACID

hypochlorous acid

Salt of an acid with 1 more oxygen atom than most common acid

NaClO₄

positive ion PER + stem + ATE

sodium perchlorate

Salt of most common acid

NaClO₃

positive ion stem + ATE

sodium chlorate

Salt of acid with 1 less oxygen atom than most common acid

NaClO₂

positive ion stem + ITE

sodium chlorite

Salt of an acid with 2 less oxygen atoms than most common acid

NaClO

positive ion HYPO + stem + ITE

sodium hypochlorite

Write the formulas for the following compounds:

3-51. magnesium chloride

3-52. phosphoric acid

3-53. boron pentachloride

3-54. ferrous chloride

3-55. carbon tetrachloride

3-56. zinc sulfide

3-57. antimony trichloride

3-58. aluminum carbonate

3-59. sulfur dichloride

3-60. aluminum nitride

3-61. lead (II) nitrate

3-62. ammonium chloride

3-63. hydrogen fluoride

3-64. hydrobromic acid

3-65. tin(IV) bromide

3-66. cuprous oxide

3-67. calcium bicarbonate

3-68. copper(I) cyanide

3-69. cesium fluoride

3-70. iron(III) phosphate

3-71. dinitrogen pentoxide

3-72. iron(II) carbonate

3-73. bromous acid

3-74. perchloric acid

3-75. magnesium cyanide

3-76. strontium carbonate

3-77. calcium nitrate

3-78. calcium sulfite

3-79. tin(IV) oxide

3-80. potassium bicarbonate

3-81. strontium chlorate

3-82. ammonium hydroxide

3-83. cadmium nitrate

3-84. diphosphorus pentoxide

3-85. sodium hydride

3-86. magnesium nitride

3-87. sulfur dioxide

3-88. aluminum nitrate

3-89. silver oxide

3-90. lithium phosphate

3-91. cuprous sulfate

3-92. ammonium fluoride

3-93. sodium sulfite

3-94. radium bicarbonate

3-95. copper(II) oxide

3-96. iron(II) sulfate

3-97. magnesium chlorate

3-98. potassium hypochlorite

3-99. disodium hydrogen phosphate

3-100. diphosphorus trisulfide

Chapter 4

BALANCING CHEMICAL EQUATIONS

Balancing Molecular Equations

Before we can begin to balance a chemical equation, we must know what the starting materials are and also what products are formed. Once we know this, the actual balancing is very simple.

A chemical equation tells us how many atoms or molecules of one substance react with how many of a second to form a certain number of products. According to the law of Conservative of Mass, *there must be the same number of atoms of each type on each side of the equation.*

In balancing any chemical equation, coefficients (numbers) may be placed before each reactant or product, as needed. That is, if we need six molecules of water, we would write 6 H_2O. We would never place a number between the symbols or alter the subscripts because so doing would change the meaning of the reaction.

Example No. 4-1. Balance: $Na_{(s)} + Cl_{2(g)} \rightarrow NaCl_{(s)}$

We note that there are 2 Cl atoms on the left side of the equation so there must be 2 on the right. Therefore, we place a 2 in front of the NaCl.

$$Na_{(s)} + Cl_{2(g)} \rightarrow 2\ NaCl_{(s)} \text{ (unbalanced)}$$

This gives us 2 Na atoms on the right side of the equation and only 1 on the left so we place a 2 in front of the Na. Now we have

$$2\ Na_{(s)} + Cl_{2(g)} \rightarrow 2\ NaCl_{(s)} \text{ (balanced)}$$

which is balanced because there are 2 Na atoms and 2 Cl atoms on each side of the equation.

Example No. 4-2. Balance: $Al_{(s)} + O_{2(g)} \rightarrow Al_2O_{3(s)}$

We see that there are 2 Al on the right side so we place a 2 in front of the Al to give 2 Al on the left side.

$$2\ Al_{(s)} + O_{2(g)} \rightarrow Al_2O_{3(s)} \text{ (unbalanced)}$$

Next, we see that there are 3 O's on the right side and O_2 on the left. To make 3 O's from O_2's we need 1½ of them, or

$$2\ Al_{(s)} + 1½\ O_{2(g)} \rightarrow Al_2O_{3(s)}$$

and now there is the same number of each type of atom on both sides of the equation. However, we know that we can never have half an atom or molecule so we double the whole equation to remove the unwanted fraction, and have

$$4\ Al_{(s)} + 3\ O_{2(g)} \rightarrow 2\ Al_2O_{3(s)} \text{ (balanced)}.$$

After you have had a little practice in balancing equations, you will be able to eliminate many of these steps or combine them with others so that the process almost becomes one of inspection, plus the use of a little patience.

Example No. 4-3. Balance: $Al(OH)_{3(aq)} + H_2SO_{4(aq)} \rightarrow Al_2(SO_4)_{3(aq)} + H_2O$

There are 2 Al on the right so we need 2 Al on the left. Thus we place a 2 in front of the $Al(OH)_3$,

$$2\ Al(OH)_{3(aq)} + H_2SO_{4(aq)} \rightarrow Al_2(SO_4)_{3(aq)} + H_2O \text{ (unbalanced)}$$

There are 3 SO_4's on the right side so we place a 3 in front of the H_2SO_4 to give 3 SO_4's on the left.

$$2\ Al(OH)_{3(aq)} + 3\ H_2SO_{4(aq)} \rightarrow Al_2(SO_4)_{3(aq)} + H_2O \text{ (unbalanced)}$$

There are 12 H's on the left side so we need 12 H's on the right and we get them by placing a 6 in front of the H_2O.

$$2\ Al(OH)_{3(aq)} + 3\ H_2SO_{4(aq)} \rightarrow Al_2(SO_4)_{3(aq)} + 6\ H_2O \text{ (balanced)}.$$

IN BALANCING, IT IS USUALLY BEST TO BEGIN WITH THE MOST COMPLEX MOLECULE, $Al_2(SO_4)_3$ in the above case.

Balance the following equations (if not already balanced). For simplicity, the indications for states have been omitted.

Combination:

4-1. $H_2 + O_2 \rightarrow H_2O$

4-2. $N_2 + O_2 \rightarrow N_2O_3$

4-3. $P + O_2 \rightarrow P_4O_6$

4-4. $K + O_2 \rightarrow K_2O$

4-5. $Fe + O_2 \rightarrow Fe_2O_3$

4-6. $Mg + N_2 \rightarrow Mg_3N_2$

4-7. $Al + Br_2 \rightarrow AlBr_3$

4-8. $S + O_2 \rightarrow SO_3$

4-9. $Li + O_2 \rightarrow Li_2O$

4-10. $Ca + O_2 \rightarrow CaO$

Decomposition:

4-11. $KClO_3 \rightarrow KCl + O_2$

4-12. $BaO_2 \rightarrow BaO + O_2$

4-13. $NH_4NO_2 \rightarrow N_2 + H_2O$

4-14. $HgO \rightarrow Hg + O_2$

4-15. $MgCO_3 \rightarrow MgO + CO_2$

4-16. $CuSO_4 \cdot 5H_2O \rightarrow CuSO_4 + H_2O$

4-17. $Pb(NO_3)_2 \rightarrow PbO + NO_2 + O_2$

4-18. $NH_3 \rightarrow N_2 + H_2$

4-19. $H_2O \rightarrow H_2 + O_2$

4-20. $(CaSO_4)_2 \cdot H_2O + H_2O \rightarrow CaSO_4 \cdot 2H_2O$

Displacement:

4-21. $MgBr_2 + Cl_2 \rightarrow MgCl_2 + Br_2$

4-22. $Al + CuSO_4 \rightarrow Al_2(SO_4)_3 + Cu$

4-23. $Zn + HCl \rightarrow ZnCl_2 + H_2$

4-24. $K + H_2O \rightarrow KOH + H_2$

4-25. $Zn + NaOH + H_2O \rightarrow Na_2Zn(OH)_4 + H_2$

4-26. $Al + H_2SO_4 \rightarrow Al_2(SO_4)_3 + H_2$

4-27. $CH_4 + Br_2 \rightarrow CHBr_3 + HBr$

4-28. $FeS_2 + O_2 \rightarrow Fe_2O_3 + SO_2$

4-29. $Al + Pb(NO_3)_2 \rightarrow Al(NO_3)_3 + Pb$

4-30. $Fe_2O_3 + H_2 \rightarrow Fe + H_2O$

Metathetical:

4-31. $Al(OH)_3 + HNO_3 \rightarrow Al(NO_3)_3 + H_2O$

4-32. $AgNO_3 + KCl \rightarrow AgCl + KNO_3$

4-33. $HgCl_2 + H_2S \rightarrow HgS + HCl$

4-34. $Na_2CO_3 + HCl \rightarrow NaCl + H_2CO_3$

4-35. $CaBr_2 + (NH_4)_2CO_3 \rightarrow CaCO_3 + NH_4Br$

4-36. $Pb(NO_3)_2 + K_2CrO_4 \rightarrow PbCrO_4 + KNO_3$

4-37. $Pb(NO_3)_2 + NaOH \rightarrow Pb(OH)_2 + NaNO_3$

4-38. $ZnCl_2 + KOH \rightarrow Zn(OH)_2 + KCl$

4-39. $(NH_4)_2SO_4 + NaOH \rightarrow NH_3 + H_2O + Na_2SO_4$

4-40. $Hg_2(NO_3)_2 + HCl \rightarrow Hg_2Cl_2 + HNO_3$

Change the following word equations into chemical symbol equations and balance:

4-41. Sodium chloride + lead nitrate $\rightarrow$ lead chloride + sodium nitrate

4-42. Barium nitrate + sulfuric acid $\rightarrow$ barium sulfate + nitric acid

4-43. Silver nitrate + potassium chromate $\rightarrow$ silver chromate + potassium nitrate

4-44. Calcium carbonate + hydrochloric acid $\rightarrow$ calcium chloride + carbon dioxide + water

4-45. Copper(II) sulfate + hydrogen sulfide $\rightarrow$ copper(II) sulfide + sulfuric acid

4-46. Iron(III) nitrate + sodium hydroxide → iron(III) hydroxide + sodium nitrate

4-47. Zinc chloride + hydrogen sulfide → zinc sulfide + hydrochloric acid

4-48. Magnesium nitrate + potassium hydroxide → magnesium hydroxide + potassium nitrate

4-49. Aluminum hydroxide + nitric acid → aluminum nitrate + water

4-50. Lead nitrate + sodium sulfate → lead sulfate + sodium nitrate

4-51. Sodium hydroxide + carbon dioxide → sodium carbonate + water

4-52. Magnesium oxide + hydrochloric acid → magnesium chloride + water

4-53. Copper + sulfur → copper(I) sulfide

4-54. Zinc + sulfuric acid → zinc sulfate + hydrogen

4-55. Iron(III) oxide + aluminum → aluminum oxide + iron

Balancing Net Ionic Reactions

Many reactions take place between ions in solutions and are indicated by the subscript (aq) meaning aqueous.

Consider the unbalanced molecular reaction

$$CaCl_2 + Na_2CO_3 \rightarrow CaCO_3 + NaCl$$

This same reaction may be written in ionic form if the reactants are placed in solution. The reaction is written as:

$$Ca^{2+}_{(aq)} + Cl^-_{(aq)} + Na^+_{(aq)} + CO_3^{2-}_{(aq)} \rightarrow CaCO_3 + Na^+_{(aq)} + Cl^-_{(aq)}$$

To simplify such equations, the (aq) designation is usually assumed and not written. So, the ionic equation becomes

$$Ca^{2+} + Cl^- + Na^+ + CO_3^{2-} \rightarrow CaCO_3 + Na^+ + Cl^-$$

Ionic reactions (reactions between ions in solution) take place when one or more of the following conditions are met:

   a. one of the products is a gas *indicated by a (g)*

   b. one of the products is an insoluble solid *indicated by an (s)*

   c. one of the products in non-ionized (*indicated by a moledular formula*)

   d. an oxidation-reduction reaction takes place (*see pages 153-157*)

Now consider the above rewritten ionic equation.

$$Ca^{2+} + Cl^- + Na^+ + CO_3^{2-} \rightarrow CaCO_{3(s)} + Na^+ + Cl^-$$

Since one of the products is a solid, the reaction will proceed.

Note that the same ions appear on both sides of the equation. Such ions, called *spectator ions*, take no part in the reaction. If we eliminate the spectator ions from this equation we have the *net ionic reaction*

$$Ca^{2+} + \cancel{Cl^-} + \cancel{Na^+} + CO_3^{2-} \rightarrow CaCO_{3(s)} + \cancel{Na^+} + \cancel{Cl^-} \text{ or}$$

$$Ca^{2+} + CO_3^{2-} \rightarrow CaCO_{3(s)}$$

Note that this net reaction is already balanced since the numbers of atoms of each type are the same on each side of the equation. Also, in an ionic equation, the sum of the charges must be the same on each side. In this case, the total charge on each side is 0.

Example No. 4-4 Balance the following ionic reaction; (aq) designations are omitted

$$H^+ + NO_3^- + K^+ + CO_3^{2-} \rightarrow CO_{2(g)} + H_2O + K^+ + NO_3^-$$

This reaction will proceed because there is a gas produced and also a nonionzed substance (H_2O).

Eliminating the spectator ions we have

$$H^+ + \cancel{NO_3^-} + \cancel{K^+} + CO_3^{2-} \rightarrow CO_{2(g)} + H_2O + \cancel{K^+} + \cancel{NO_3^-}$$

or

$$H^+ + CO_3^{2-} \rightarrow CO_{2(g)} + H_2O$$

To balance the above net ionic reaction we note that there are 2 H's on the right side of the equation so we place a 2 in front of the H^+. The number of carbons and oxygens is already the same on each side of the equation. The ent balanced ionic equation becomes

$$2H^+ + CO_3^{2-} \rightarrow CO_{2(g)} + H_2O$$

This equation is balanced because the number of atoms of each type is the same on each side of the equation and also the net charge is the same on each side.

Give the balanced net ionic equation for the following reactions:

Ionic:

4-56. $Na^+ + Cl^- + Pb^{2+} + NO_3^- \rightarrow PbCl_{2(s)} + Na^+ + NO_3^-$

4-57. $Ba^{2+} + Cl^- + K^+ + CO_3^{2-} \rightarrow BaCO_{3(s)} + K^+ + Cl^-$

4-58. $Al^{3+} + NO_3^- + Na^+ + OH^- \rightarrow Al(OH)_{3(s)} + Na^+ + NO_3^-$

4-59. $Ag^+ + NO_3^- + Zn^{2+} + Cl^- \rightarrow AgCl_{(s)} + Zn^{2+} + NO_3^-$

4-60. $Ca^{2+} + Cl^- + K^+ + C_2O_4^{2-} \rightarrow CaC_2O_{4(s)} + K^+ + Cl^-$

4-61. $Mg^{2+} + NO_3^- + Na^+ + OH^- \rightarrow Mg(OH)_{2(s)} + Na^+ + NO_3^-$

4-62. $Fe^{3+} + Cl^- + Na^+ + OH^- \rightarrow Fe(OH)_{3(s)} + Na^+ + Cl^-$

4-63. $Na^+ + C_2H_3O_2^- + H^+ + Cl^- \rightarrow Na^+ + Cl^- + HC_2H_3O_2$

4-64. $K^+ + OH^- + H^+ + SO_4^{2-} \rightarrow H_2O + K^+ + SO_4^{2-}$

4-65. $Na^+ + SO_3^{2-} + H^+ + Cl^- \rightarrow Na^+ + Cl^- + H_2O + SO_{2\,(g)}$

4-66. $MgO_{(s)} + H^+ + Cl^- \rightarrow Mg^{2+} + Cl^- + H_2O$

4-67 $Ca^{2+} + OH^- + Li^+ + PO_4^{3-} \rightarrow Ca_3(PO_4)_{2(s)} + Li^+ + OH^-$

4-68. $Ba^{2+} + Cl^- + H^+ + SO_4^{2-} \rightarrow BaSO_{4(s)} + H^+ + Cl^-$

4-69. $H_2S_{(g)} + Pb^{2+} + (C_2H_3O_2)^- \rightarrow PbS_{(s)} + HC_2H_3O_2$

4-70. $Mg(OH)_{2(s)} + H^+ + NO_3^- \rightarrow H_2O + Mg^{2+} + NO_3^-$

Chapter 5

ATOMIC AND MOLECULAR MASSES
(as determined from the table of atomic masses)

ATOMIC MASSES. The mass of the Carbon-12 atom at 12.0000 units has been adopted as the official standard for atomic masses. All other atomic masses are based on these same units. Hence when we say that the atomic mass of Helium is 4.00260, we are saying that Helium has a mass 4.00260/12.0000 of the Carbon-12 atom. Another way of saying the same thing is to say that when we use such units that 12.0000 of them are required to equal the mass of the Carbon-12 atom, the Helium atom will have a mass of 4.00260 of these same units.

The MOLECULAR MASS (FORMULA MASS) of a compound is the sum of the atomic masses of all of the elements present in the molecule. Thus, H_2O contains 2 atoms of Hydrogen (atomic mass 1.008) and 1 atom of Oxygen (atomic mass 15.999) so that the molecular mass (to 3 decimal places) is

$$(2 \times 1.008) + (1 \times 15.999) = 18.015$$

The MOLE, abbreviated mol, contains as many atoms, molecules, or other particles as there are atoms in 0.012 kg of carbon-12. A mole is equivalent to the atomic mass or the molecular mass expressed in grams.

Thus,   1 mole of oxygen atoms, O, weighs 15.9994 g;

1 mole of oxygen molecules, O_2, weighs 2 × 15.9994 g or 31.9998 g;

1 mole methyl alcohol, CH_3OH, weighs

12.011 g + 3 × 1.008 g + 15.9994 g + 1.008 g or 32.042 g.

5 moles methyl alcohol weigh 5 × 32.042 g or 160.210 g.

The mass of one mole of compound is known as the MOLAR MASS of that compound.

Thus, the molecular mass of water is 18.015

and the molar mass of water is 18.015 g.

That is, the mass of one mole of water is 18.015 g.

Example No. 5-1.   Determine the molecular mass of sulfuric acid, H_2SO_4 (to 2 decimal places).

2 H = 2 × 1.01 = 2.02
1 S = 1 × 32.07 = 32.07
4 O = 4 × 16.00 = 64.00

Molecular Mass H_2SO_4 = 98.07

Example No. 5-2.   Determine the molar mass of lead nitrate $Pb(NO_3)_2$ (to 1 decimal place).

1 Pb = 1 × 207.2 = 207.2 g
2 N = 2 × 14.1 = 28.2 g
6 O = 6 × 16.0 = 96.0 g

Molar Mass $Pb(NO_3)_2$ = 331.2 g

Example No. 5-3. How many moles are contained in 8.00 g NaOH? The molar mass of NaOH is

$$1 \text{ Na} = 1 \times 22.990 \text{ g}$$

$$1 \text{ O} = 1 \times 15.999 \text{ g}$$

$$\underline{1 \text{ H} = 1 \times \ \ 1.0079 \text{ g}}$$

Molar Mass NaOH = 39.9969 g = 40.0 g (3 significant figures)

Therefore, 8.00 g NaOH is $\dfrac{8.00 \text{ g}}{40.0 \text{ g/mole}}$ = 0.200 moles

Example No. 5-4. Determine the molecular mass of $BaCl_2$ $2H_2O$ (to 2 decimal places).

$$1 \text{ Ba} = 1 \times 137.33 = 137.33$$

$$2 \text{ Cl} = 2 \times 35.45 = \ \ 70.90$$

$$4 \text{ H} = 4 \times \ \ 1.01 = \ \ \ \ 4.04$$

$$\underline{2 \text{ O} = 2 \times \ \ 16.00 = \ \ \ 32.00}$$

Molecular Mass $BaCl_2$ $2H_2O$ = 244.27

Example No. 5-5. How many moles are there in 5.00 g Aluminum? The molar mass of aluminum is 26.98 g, so

5.00 g Al is $\dfrac{5.00 \text{ g}}{26.98 \text{ g/mole}}$ = 0.185 moles.

Refer to Chapter 6 for other methods of determining atomic and molecular masses.

Problem Assignment

Determine the molecular masses of the following compounds (to 2 decimal places):

5-1.	MgO	(40.30)	5-11.	MnO_2	
5-2.	$CaSO_4$	(136.15)	5-12.	K_2HPO_4	
5-3.	Fe_2O_3	(159.70)	5-13.	K_2PtCl_6	
5-4.	$(NH_4)_2Cr_2O_7$	(252.10)	5-14.	$NaClO_4$	
5-5.	$K_2Cr_2O_7$	(294.18)	5-15.	P_4O_{10}	
5-6.	$CuSO_4$ $5H_2O$		5-16.	$C_{12}H_{22}O_{11}$	
5-7.	$KC_2H_3O_2$		5-17.	$(CaSO_4)_2$ H_2O	
5-8.	$Fe_2(SO_4)_3$		5-18.	$K_3Fe(CN)_6$	
5-9.	CI_4		5-19.	$Ag(NH_3)_2Cl$	
5-10.	NH_4Cl		5-20.	$NaAl(SO_4)_2$ $12H_2O$	

How many moles are contained in:

5-21.	3.75 g NaCl	(0.0643)	5-26.	5.04 g Na	
5-22.	2.45 g H_2SO_4		5-27.	39.6 g S	
5-23.	4.60 g CO_2		5-28.	12.5 g Ca	
5-24.	13.5 g $MgSO_4$		5-29.	9.62×10^{-2} g He	
5-25.	120 g CH_4		5-30.	0.415 lb. Al	

Chapter 6

THE KINETIC MOLECULAR THEORY AND THE GAS LAWS

I. THE KINETIC MOLECULAR THEORY

The fact that all gases behave similarly with a change in temperature and pressure led to the Kinetic-Molecular Hypothesis which, after much confirmatory evidence, became the Kinetic Molecular Theory. The main assumptions of the theory are:

A. All matter is composed of discrete particles (atoms or molecules).
B. In gases the particles are relatively far apart.
C. The particles of all substances are in continuous rapid motion.
D. Collisions between particles are perfectly elastic. (That is, when particles collide with one another or with the walls of the container, they rebound without any loss in energy. This assumption is necessary to account for the fact that the pressure exerted by a gas does not vary after a long period of time. In other words, the particles must move perpetually and rebound from collisions with no loss in energy to account for these facts.)
E. The average kinetic energy of the particles is directly proportional to the Kelvin temperature.

Let us see if we can explain the experimental facts on the basis of these assumptions.

(a) *Gases exert pressure.* The pressure exerted by a gas is due to the particles striking the wall of the container. Each particle that strikes the wall rebounds without any loss in energy and when doing so exerts a pressure on the wall. While the force exerted by one particle is very small, the number of collisions per second on any given area is enormous. Since the particles are in rapid motion and strike one another as well as the container, they will be moving in all possible directions and so exert the same pressure over all the container walls.

(b) *Gases are highly compressible.* The particles of a gas are relatively far apart and there is a great deal of empty space between them. The rapid motion of the particles makes them appear to occupy the space but at any given instant the majority of the space is empty. Therefore, compression crowds the particles closer together, decreasing the space between them. Conversely, decreasing the pressure exerted on a gas allows the particles to move farther apart and occupy more space.

(c) *Gases diffuse very easily.* The ability of gases to diffuse into any empty container or into another gas is due to the rapid motion of the particles and to the ability of the empty space between the particles to receive a new type of particle. That is, any gas consists of rapidly moving particles separated by great distances from one another. These particles move in straight lines and collide with one another as well as with the walls of the container. As a result of these collisions, the particles are moving in all possible directions and thus diffuse throughout the container. It makes no difference whether the container is empty or holds another gas since even in another gas the majority of the space is empty.

(d) *Pressure exerted by a gas increases with the temperature if the volume remains constant.* As the temperature rises, there is an increase in the kinetic energy of the particles due to an increase in their velocity. These more rapidly moving particles will strike the walls of the container with increased energy and so exert a greater pressure. Also there will be more impacts per second due to the increased velocity.

(e) *Gases expand on heating (if the pressure remains constant).* An increase in temperature will increase the velocity of the particles as stated above. If the volume remained fixed, the pressure would increase due to the greater energy of the particles striking the walls of the container. However, since we are keeping the pressure constant, the only way we can do this is to allow the gas to expand and thus reduce the number of collisions with the wall so as to maintain the same pressure.

II. UNITS OF MEASUREMENT FOR GASES*

When we are talking about a gas, we usually want to know its pressure, temperature, and volume. While there are many methods for expressing pressure, the most common units are the atmosphere, pascals (Pa), and torr. Temperature is measured in °C or K and volume in cm³, mL, or L.

Pressure is measured by a barometer, one form of which is shown at the right. A glass tube about 1 yard long is filled with mercury and the open end inverted into a dish of mercury. The mercury falls in the tube, leaving a vacuum above it, until the pressure exerted by the air just balances the mercury column in the tube. Thus the atmospheric pressure is expressed as being equal to so many millimeters of mercury or the height of the column from A to B.

One unit of pressure is the torr, named after the Italian scientist Torricelli, who invented the barometer. One torr is the pressure exerted by one millimeter of mercury at sea level.

We must have a common standard for measurement before we can compare **gases** under different conditions of temperature and pressure. STP stands for standard temperature and pressure. Standard temperature is 0°C or 273 K and standard pressure is 1.00 atmosphere. Standard pressure may be expressed in many different units. Among these are:

760 torr	29.92 in. Hg	1.013×10^5 pascals (Pa)
760 mm Hg	14.7 lb./in.²	101.3 kilopascals (kPa)

III. THE GAS LAWS

A. Boyle's Law

AT CONSTANT TEMPERATURE, THE VOLUME OCCUPIED BY A GIVEN MASS OF GAS IS INVERSELY PROPORTIONAL TO THE PRESSURE, or, as we can say in simpler terms, as the pressure goes up, the volume of a gas goes down and as the pressure goes down, the volume goes up. We can express this algebraically as

$$\frac{V_1}{V_2} = \frac{P_2}{P_1} \text{ or } P_1V_1 = P_2V_2$$

where V_1 is the initial volume and V_2 the final volume, P_1 the initial pressure and P_2 the final pressure. We can solve Boyle's Law problems either by the use of the equation given above or by the use of logical reasoning based upon the same ideas. The latter is the better method since it teaches us to think. Let us try a few problems to compare these two methods and also to see how Boyle's Law may be applied in various types of problems.

Example No. 6-1.   A container holds 500 mL of a gas at 20°C and 735 torr. What will be the volume of the gas if the pressure is increased to 787 torr (temperature remaining constant)?

method (1)—We can use the formula $P_1V_1 = P_2V_2$ and get

735 torr × 500 mL = 787 torr × V_2.

so that $V_2 = \dfrac{735 \text{ torr} \times 500 \text{ mL}}{787 \text{ torr}} = 467$ mL

*The International Union of Pure and Applied Chemistry (IUPAC) has defined standard pressure for reporting thermodynamic data as 1 bar. 1 bar = 100,000 pascals (10^5 kPa) and is equal to 0.987 atm. Since the bar and the atmosphere differ by a very small percentage, we will continue to use the atmosphere as the standard of pressure in this book.

method (2)—Since the volume is inversely proportional to the pressure, as the pressure goes up, the volume goes down. Therefore, we multiply the original volume, 500 mL, by the ratio of the pressures. Since the volume must go down, we want the smaller pressure on top or we have

$$500 \text{ mL} \times \frac{735 \text{ torr}}{787 \text{ torr}} = 467 \text{ mL}.$$

This method is to be preferred since we should always be able to reason out a problem, while we may forget the formula which applies to that specific case.

Example No. 6-2.   A gas tank holds 2500 liters at 820 torr. If the temperature remains constant, what will be the volume of the gas at standard pressure?

method (1)—$P_1V_1 = P_2V_2$

$$820 \text{ torr} \times 2500 \text{ liters} = 760 \text{ torr} \times V_2$$
$$V_2 = 2500 \text{ liters} \times \frac{820 \text{ torr}}{760 \text{ torr}} = 2700 \text{ liters}$$

method (2)—Since the pressure is going down, the volume must go up and we want the larger pressure on top so that the volume will increase. Thus we have

$$\text{Volume} = 2500 \text{ liters} \times \frac{820 \text{ torr}}{760 \text{ torr}} = 2700 \text{ liters}$$

Example No. 6-3.   A cylinder contains 5.0 liters of O_2 at a pressure of 10 atmospheres. If the temperature remains constant, what will be the volume when the pressure is reduced to 2.0 atmospheres?

Although the pressure here is given in atmospheres, this does not mean that we have to convert to mm Hg before we can work the problem. We can use any units of pressure as long as we use the same units throughout any given problem because the units themselves will always cancel. Thus we have

$$P_1V_1 = P_2V_2 \text{ and } 10 \text{ atm} \times 5.0 \text{ liters} = 2.0 \text{ atm} \times V_2$$
$$V_2 = 25 \text{ liters}$$

or, since the pressure is going down, the volume must go up so we place the larger pressure (10 atm) on top and the lower pressure (2 atm) on the bottom and we have

$$\text{Volume} = 5.0 \text{ liters} \times \frac{10 \text{ atm}}{2.0 \text{ atm}} = 25 \text{ liters}$$

B. Charles' Law

AT CONSTANT PRESSURE, THE VOLUME OF A GAS IS DIRECTLY PROPORTIONAL TO ITS KELVIN OR ABSOLUTE TEMPERATURE. The word Kelvin (or Absolute) is very important in this definition and should always be included. Charles, a French Scientist, found that the volume of a gas at 0°C increased by 1/273 for each degree rise in its temperature and decreased 1/273 for each degree that its temperature fell. Thus, if the gas were cooled 1°C its volume would decrease by 1/273 and if it were cooled 273°C its volume would decrease by 273/273 or would be zero. However, the gas would liquify before it reached this temperature, which he called absolute zero.

Algebraically, Charles' Law can be stated as

$$\frac{V_1}{V_2} = \frac{T_1}{T_2}$$

where V_1 and V_2 are the initial and final volumes and T_1 and T_2 the initial and final Kelvin temperatures. We shall use the logical method for solving these problems as we did with Boyle's Law.

Example No. 6-4.   50.0 mL of a gas at 25°C will occupy what volume if heated to 35°C (pressure constant)?

First we must change both of these temperatures to Kelvin; 25°C = 298 K and 35°C = 308 K. Since the temperature is going up, the volume must also go up and therefore, we want the larger Kelvin temperature on top, or

$$\text{Volume} = 50.0 \text{ mL} \times \frac{308 \text{ K}}{298 \text{ K}} = 51.6 \text{ mL}$$

Example No. 6-5.   A sample of gas has a volume of 250 mL at 50°C. What is its volume at standard temperature (0°C) if the pressure remains constant?

First, 50°C = 323 K and 0°C = 273 K. Since the temperature is going down, the volume is going down and we want the smaller Kelvin temperature on top. Thus we have

$$\text{New Volume} = 250 \text{ mL} \times \frac{273 \text{ K}}{323 \text{ K}} = 211 \text{ mL}$$

C. Combination of Boyle's and Charles' Law

It is sometimes very difficult to maintain constant temperature and pressure in the laboratory so that we generally allow both to vary and calculate the results with a combination of gas laws. Also, this combination is useful in calculating volumes to STP as is required in many problems.

We can work problems by using the formula $\dfrac{P_1 V_1}{T_1} = \dfrac{P_2 V_2}{T_2}$

or by logical reasoning. Again, we shall use the latter method. When calculating the results of a change in pressure and temperature, we work the problem just as if it were two separate problems with first, a change in pressure with temperature constant and second, a change in temperature with pressure constant.

Example No. 6-6.   A certain gas occupies 500 mL at 30°C and 720 torr. What will be its volume at STP?

Let us consider the pressure first; it is going up from 720 torr to 760 torr and, therefore, the volume must go down. So we place the smaller pressure on top, or

$$\text{Volume} = 500 \text{ mL} \times \frac{720 \text{ torr}}{760 \text{ torr}}$$

Next we consider the temperature—it is going down from 30°C to 0°C or from 303 K to 273 K so that the volume will also go down. Therefore, we want the smaller temperature on top, or the volume for both changes will be

$$\text{New Volume} = 500 \text{ mL} \times \frac{720 \text{ torr}}{760 \text{ torr}} \times \frac{273 \text{ K}}{303 \text{ K}} = 427 \text{ mL}$$

Example No. 6-7.   What volume will 5.0 liters of gas at −40°C and 700 torr occupy if the temperature is changed to 30°C and the pressure to 800 torr?

Since the pressure is going up, the volume must go down and so we place the smaller pressure on top, or

$$\text{Volume} = 5.0 \text{ liters} \times \frac{700 \text{ torr}}{800 \text{ torr}}$$

Since the temperature is going up from −40°C to 30°C or from 233 K to 303 K, the volume must also go up, so we place the larger temperature on top. Therefore,

$$\text{New Volume} = 5.0 \text{ liters} \times \frac{700 \text{ torr}}{800 \text{ torr}} \times \frac{303 \text{ K}}{233 \text{ K}} = 5.7 \text{ liters}$$

We can obtain the same results by using a formula, but formulas can be forgotten, while we should always be able to reason logically.

D. Dalton's Law

It is not always possible to collect a gas in the pure state. That is, it may be mixed with other gases. For instance, in the collection of H_2 by the downward displacement of water (see diagram—a) the gas will always contain some water vapor. In the collection of Cl_2 by upward displacement of air (see diagram—b) the Cl_2 will be mixed with air. Dalton's Law states—IN A MIXTURE OF GASES, EACH GAS EXERTS A PRESSURE INDEPENDENT OF OTHERS AND PROPORTIONAL TO THE RELATIVE AMOUNT BY VOLUME OF THAT GAS IN THE MIXTURE. Thus, a mixture of gases which is 80% by volume N_2 and 20% by volume O_2 at a total atmospheric pressure of 760 torr will have 80% of 760 torr or 608 torr due to the N_2 and 20% of 760 torr or 152 torr due to the O_2. Each gas exerts its own pressure as if it alone were present, since in the gaseous state the molecules are so far apart that there is plenty of room for other types of particles to exist without interaction between them. Thus, if we collect a gas over water, we will have a mixture of two gases—the one collected plus some water vapor. If we knew the concentration or the amount by volume of the water vapor, we could calculate its pressure, which we call the partial pressure since it is due to only part of the gases present. It would be difficult and tedious to measure the amount of water vapor present in a sample collected over water, but fortunately we have another means of obtaining the vapor pressure of the water. We get this from the following chart, since it has been found that the pressure due to the water vapor is dependent only upon the temperature and not upon the size of the container. Therefore, if we know the temperature, we can find the partial pressure due to the water vapor by referring to the following chart.

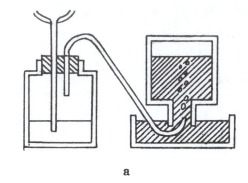

a

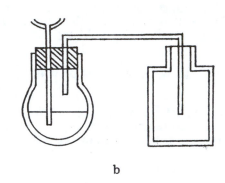

b

VAPOR PRESSURE OF WATER AT VARIOUS TEMPERATURES

Temp °C	Pressure torr	Pressure kPa × 10³	Temp °C	Pressure torr	Pressure kPa × 10³	Temp °C	Pressure torr	Pressure kPa × 10³
0	4.6	0.61	21	18.6	2.48	30	31.8	4.24
5	6.5	0.87	22	19.8	2.64	40	58.3	7.38
10	9.2	1.23	23	21.1	2.81	50	92.5	12.3
15	12.8	1.70	24	22.4	2.99	60	149.4	19.93
16	13.6	1.81	25	23.8	3.17	70	233.7	31.18
17	14.5	1.93	26	25.2	3.36	80	355.1	47.37
18	15.5	2.07	27	26.7	3.56	90	525.8	70.12
19	16.5	2.20	28	28.3	3.77	100	760	101.3
20	17.5	2.34	29	30.0	4.00			

Whenever we collect a gas over water, we always adjust the receiver so that the level of the water inside is equal to the level outside of the receiver (diagram 1) making the total pressure equal to atmospheric pressure.

If the gas level in the receiver is higher than that of the container, the pressure of the gas is less than atmospheric presusre, and if the level is lower, it is more than atmospheric pressure. In either case we would have to measure the difference in heights and correct for the pressure due to a column of water that high. This involves a great deal of unnecessary work and calculations, so we always adjust the levels to make them equal before taking a set of readings.

<u>Diagram 1</u>

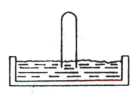 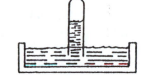 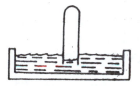

P gas = P atm P gas < P atm P gas > P atm

Example No. 6-8. A sample of H_2 collected over water at 25°C and 760 torr occupies 100 mL. What volume will the gas occupy at 25°C and 760 torr dry?

We note that the temperature remains constant so that all we have to do is to correct for the change in pressure. We assume that the levels are equal or it would have been stated in the problem so we can draw the following diagram. The pressure exerted by the gas must just equal atmospheric pressure to keep the levels the same. Thus we can write:

P atm = 760 torr P gas = 760 torr

Pressure H_2 + Pressure H_2O = Pressure atm = 760 torr

However, from our table we find the pressure of the water vapor at 25°C is 23.8 torr, so

Pressure H_2 + 23.8 torr = 760 torr and Pressure H_2 = 760 torr – 23.8 torr = 736.2 torr, so 736.2 torr is the partial pressure of the H_2. When we calculate the volume dry at 760 torr, we see that the pressure is going up so that the volume must go down and thus we place the smaller number on top—

Volume dry gas at 25°C and 760 torr = 100 mL × $\dfrac{736.2 \text{ torr}}{760 \text{ torr}}$ = 96.8 mL

E. Combination of Boyle's, Charles', and Dalton's Laws

Example No. 6-9. The volume of a sample of gas collected over water is 453 mL at 18°C and 780 torr. What is its volume dry at STP?

Since the pressure 780 torr is due to the gas plus water vapor, we can find the pressure of the gas alone by subtracting the water vapor pressure from 780 torr. Thus,

P_{gas} + $P_{water\ vapor}$ = 780 torr and

P_{gas} = 780 torr - $P_{water\ vapor}$

The water vapor pressure at 18°C is 15.5 torr so that the pressure of the gas is

780 torr – 15.5 torr = 764.5 torr.

The pressure is going down from 764.5 torr to 760 torr so that the volume is going up and we place the larger pressure on top. Since the temperature is going down, from 18°C to 0°C or from 291 K to 273 K, the volume is also going down and we place the smaller temperature on top.

New Volume = 453 mL × $\dfrac{764.5 \text{ torr}}{760 \text{ torr}}$ × $\dfrac{273 \text{ K}}{291 \text{ K}}$ = 429 mL

Example No. 6-10. A gas is collected over water and occupies 517 mL at 50°C and 800 torr. What will be its volume, dry, at 30°C and 600 torr?

At 50°C the vapor pressure is 92.5 torr so that the partial pressure of the gas is 800 torr – 92.5 torr = 707.5 torr. Since the pressure is going down, from 707.5 torr to 600 torr, the volume is going up and we place the larger pressure on top. Since the temperature is going down from 50°C to 30°C, or from 323 K to 303 K, the volume is going down, and we place the smaller temperature on top. Thus,

$$\text{Volume} = 517 \text{ mL} \times \frac{707.5 \text{ torr}}{600 \text{ torr}} \times \frac{303 \text{ K}}{323 \text{ K}} = 572 \text{ mL}$$

Example No. 6-11. A 250 mL sample of a dry gas at 30°C and 748 torr is transferred to a container over water at the same temperature and pressure. What will be the new volume of the gas?

The partial pressure of water at 30°C is 31.8 torr so that the partial pressure of the gas over water will be 748 torr – 31.8 torr = 716.2 torr. Since the pressure is going down, from 748 torr to 716.2 torr, the volume must be going up, and since the temperature remains constant, we have

$$\text{Volume} = 250 \text{ mL} \times \frac{748 \text{ torr}}{716.2 \text{ torr}} = 261 \text{ mL}$$

Example No. 6-12. A 500 mL sample of gas, dry, at STP is transferred to a container over water at 15°C and 740 torr. What will be the new volume of the gas?

The vapor pressure of water at 15°C is 12.8 torr so that the partial pressure of the gas is 740 torr – 12.8 torr or 727.2 torr. Since the pressure is going down from 760 torr to 727.2 torr, the volume will go up, and since the temperature is increasing, the volume will again go up, so we have:

$$\text{Volume} = 500 \text{ mL} \times \frac{760 \text{ torr}}{727.2 \text{ torr}} \times \frac{288 \text{ K}}{273 \text{ K}} = 551 \text{ mL}$$

F. The General Gas Law

The formula for both Charles' and Boyle's Laws

$$\frac{P_1 V_1}{T_1} = \frac{P_2 V_2}{T_2}$$

may also be written as $PV/T = k$ or $PV = kT$.

The constant k includes the number of moles present. The general gas law is given as

$$\boxed{PV = nRT}$$

where R is the gas constant and n is the number of moles present. The value of R depends upon the units selected for P and V. T is always Kelvin.

Values of R the gas constant

Units of V	Units of P	R
liters	atmospheres	0.08206 L·atm/mol·K
liters	torr	62.364 L·torr/mol·K
liters	cm of mercury	6.236 L·cm/mol·K

other values of R are: $1.987 \dfrac{\text{cal}}{\text{mol·K}}$ and $8.31 \dfrac{\text{kPa·dm}^3}{\text{mol·K}}$

Example No. 6-13. Calculate the volume in liters occupied by 0.5000 mol of a gas at 286.0°C and 2.500 atm pressure.

Using the formula $PV = nRT$, $V = nRT/P$

Then, selecting the value of R with the proper units, and also changing 286.0°C to Kelvin

$$V = \frac{nRT}{P} = \frac{0.5000 \text{ mol} \times 0.08206 \frac{\text{L·atm}}{\text{mol·K}} \times 559.0 \text{ K}}{2.500 \text{ atm}}$$

$V = 9.176$ liters

Example No. 6-14. 0.975 grams of a gas at 180°C and 505 torr pressure occupy a volume of 0.507 liters. Calculate the molecular mass of the gas.

Using $PV = nRT$, $n = PV/RT$. Selecting the proper units for R,

$$n = \frac{PV}{RT} = \frac{50.5 \text{ cm} \times 0.507 \text{ L}}{62.364 \frac{\text{L·torr}}{\text{mol·K}} \times 453 \text{ K}} = 0.00906 \text{ mol}$$

Then, 0.00906 moles of the gas weigh 0.975 g, or $\frac{0.975 \text{ g}}{0.00906 \text{ mol}}$ which equal 108 g/mol. Thus, the molecular mass of the gas is 108.

G. Deviation of Real Gases from Perfect Gas Behavior

The general gas equation $PV = nRT$ represents the behavior of an ideal gas. It also represents the behavior of almost all gases under ordinary conditions of temperature and pressure.

However, under conditions of high pressure or low temperature gases do not behave ideally. There are two reasons for deviations from ideal behavior. First, the particles themselves have a definite size and so occupy an appreciable part of the volume under high pressure. This causes the volume of the gas to be greater than that calculated for an ideal gas. Second, particles are brought closer together, hence attract each other more strongly especially under conditions of high pressure and low temperature. This attraction causes the volume of the gas to be smaller than that calculated for an ideal gas.

The van der Waals' equation corrects for these deviations. It is given as:

$$\left(P + \frac{n^2 a}{V^2}\right)(V - nb) = nRT$$

where a and b are constants depending upon the gas being considered.

For carbon dioxide, $a = 3.59$ L^2 atm/mol^2 and $b = 0.0423$ L/mol.

Example No. 6-15. 1.00 mole of carbon dioxide gas occupies a volume of 0.250 liters at 60°C. Calculate the pressure of the gas (a) assuming that CO_2 is an ideal gas and (b) using van der Waals' equation.

(a) Using $PV = nRT$, $P = nRT/V$

$$P = \frac{1.00 \text{ mol} \times 0.08206 \text{ L·atm/mol·K} \times 333 \text{ K}}{0.250 \text{ L}} = 109 \text{ atm}$$

(b) Using $\left(P+\dfrac{n^2a}{V^2}\right)(V-nb)=nRT$

$$P=\left(\frac{nRT}{V-nb}\right)-\frac{n^2a}{V^2}$$

$$P=\left(\frac{1.00\ \text{mol}\times0.08206\dfrac{\text{L·atm}}{\text{mol·K}}\times333\ \text{K}}{0.250\ \text{L}-1.00\ \text{mol}\times0.0423\ \dfrac{\text{L}}{\text{mol}}}\right)-\left(\frac{(1.00\ \text{mol})^2\times3.59\dfrac{\text{L}^2\text{·atm}}{\text{mol}^2}}{(0.250\ \text{L})^2}\right)=$$

$=131.9\text{atm} - 59.4\ \text{atm} = 72.5\ \text{atm}.$

Let us see if these deviations are of any consequence at ordinary temperatures and pressures.

Example No. 6-16. 1.00 mole of oxygen occupies a volume of 20.0 liters under a pressure of 1.50 atmospheres. Calculate the temperature of this gas (a) using the general gas law and (b) using van der Waals' equation. For oxygen, $a=1.36$ L^2·atm/mole^2 and $b=0.0318$ L/mole.

(a) Using $PV = nRT$, $T = PV/nR$

$$T = \frac{1.50\ \text{atm} \times 20.0\ \text{L}}{1.00\ \text{mol} \times 0.08206\ \dfrac{\text{L·atm}}{\text{mol·K}}} = 366\ \text{K}$$

(b) Using $\left(P+\dfrac{n^2a}{V^2}\right)(V-nb)= nRT$

$$T = \frac{\left(P+\dfrac{n^2a}{V^2}\right)(V-nb)}{nR}$$

$$T = \left(\frac{\left(1.50\ \text{atm} + \dfrac{(1\ \text{mol})^2\ (1.36\ \text{L}^2\text{·atm/mol}^2)}{(20.0\ \text{L})^2}\right)\left(20.0\ \text{L}-(1\ \text{mol})\ (0.0318\ \dfrac{\text{L}}{\text{mol}})\right)}{1\ \text{mol}\ (0.08206\ \text{L·atm/mol·K})}\right)$$

$T = 366\ \text{K}$

H. Graham's Law of Diffusion

We are all familiar with the fact that gases diffuse through the air, some very rapidly, others more slowly. A bottle of perfume opened in one corner of the room can soon be detected in all parts of the room. A small amount of Cl_2 gas introduced into a bottle will soon show the yellow color of the gas distributed uniformly throughout the container.

When we compare the rates of diffusion of two gases, we would expect the one with the heavier molecules to diffuse much slower than the one with the lighter molecules. This was stated quantitatively by Graham as follows: The rates of diffusion of gases are inversely proportional to the square roots of their densities, or

$$\frac{\text{rate of gas 1}}{\text{rate of gas 2}} = \frac{\sqrt{\text{density of gas 2}}}{\sqrt{\text{density of gas 1}}}$$

Densities of gases are not always easy to obtain, but if we remember that $D = M/V$ where the mass is the molar mass and the volume is the molar volume we have

$$\frac{\text{rate of gas 1}}{\text{rate of gas 2}} = \frac{\sqrt{\dfrac{\text{molar mass gas 2}}{\text{molar volume gas 2}}}}{\sqrt{\dfrac{\text{molar mass gas 1}}{\text{molar volume gas 1}}}}$$

As we shall see later in this chapter, the molar volumes of all gases are equal under the same conditions of temperature and pressure so we have

$$\frac{\text{rate of gas 1}}{\text{rate of gas 2}} = \frac{\sqrt{\text{molar mass of gas 2}}}{\sqrt{\text{molar mass of gas 1}}}$$

which is in a somewhat more useable form. From this we can determine the relative rates of diffusion of gases if we know their molar masses or their molecular formulas; if we know the rates of diffusion, we can calculate the molar mass of any gas from the molar mass of a standard gas.

Example No. 6-17.   What are the relative rates of diffusion of H_2 and O_2?

The molar mass of H_2 is 2.0 and O_2, 32. From Graham's Law we have

$$\frac{\text{rate } H_2}{\text{rate } O_2} = \frac{\sqrt{\text{molar mass of } O_2}}{\sqrt{\text{molar mass of } H_2}} = \sqrt{\frac{32}{2.0}} = \frac{5.6}{1.4} = \frac{4.0}{1.0}$$

or H_2 will diffuse 4 times as fast as O_2. If the rate of diffusion of H_2 is 50.0 ft./sec., the rate of O_2 will be ¼ of that or 12.5 ft./sec. or if H_2 escapes from an orifice at the rate of 800 cm³/sec., O_2 will escape from the same orifice at the rate of 200 cm³/sec.

Example No. 6-18.   List the following gases in order of increasing rates of diffusion.

$$H_2, N_2, Ne, Cl_2, HCl, CH_4, He.$$

Since all we want is the order of increasing rates of a diffusion and not the relative rates, all we have to do is arrange them in order of decreasing molar masses, because the one with the greatest molar mass will diffuse the slowest and the one with the smallest molar mass will diffuse the fastest. So we list the substances and their molar masses as follows:

H_2	N_2	Ne	Cl_2	HCl	CH_4	He
2	28	20	71	36.5	16	4

and, therefore, the order of increasing rates of diffusion is:

Cl_2	HCl	N_2	Ne	CH_4	He	H_2
71	36.5	28	20	16	4	2

Example No. 6-19.   What are the relative rates of diffusion of N_2 and Cl_2?

$$\frac{\text{rate } N_2}{\text{rate } Cl_2} = \frac{\sqrt{\text{molar mass of } Cl_2}}{\sqrt{\text{molar mass of } N_2}} = \frac{\sqrt{71}}{\sqrt{28}} = \frac{8.43}{5.3} = \frac{1.6}{1}$$

or N_2 will diffuse 1.6 times as fast as Cl_2.

IV. MOLAR VOLUMES

Since equal volumes of gases at the same temperature and pressure contain the same number of molecules, and since the volume of each different gas has a different mass, then the masses of the molecules themselves must be different.

The mass of 22.4 liters of a gas at STP is called the molar mass of the gas and this volume is called the molar volume. This number was selected because it is the volume occupied by 32 g of O_2 at STP and because O_2 was the standard for comparison in the calculation of molecular masses.

The number of molecules actually present in this molar volume has been calculated by several different methods and found to be 6.02×10^{23} (which is called Avogadro's number or N).

THIS NUMBER REPRESENTS THE NUMBER OF MOLECULES IN 1 MOL OF ANY COMPOUND.

Example No. 6-20. How many molecules are there in 1.00 liter of oxygen at 0°C and 1.00×10^{-5} torr?

We first change liters O_2 at STP to mol O_2 at STP.

$$1.00 \ \cancel{L \ O_2} \times \frac{1 \ mol \ O_2}{22.4 \ \cancel{L \ O_2}} = 0.0446 \ mol \ O_2 \ at \ STP$$

Then, changing mol O_2 to molecules O_2 since 1 mol contains Avogadro's number or 6.02×10^{23} molecules we have

$$0.0446 \ \cancel{mol \ O_2} \times \frac{6.02 \times 10^{23} \ molecules \ O_2}{1 \ \cancel{mol \ O_2}} = 2.68 \times 10^{22} \ molecules \ O_2.$$

Thus we have 2.68×10^{22} molecules O_2 at STP, 0°C and 760 torr, or 2.68×10^{22} molecules O_2 at 760 torr or 2.68×10^{22} molecules O_2/760 torr.

For a pressure of 10^{-5} torr, we have $\dfrac{2.68 \times 10^{22} \ molecules \ O_2}{760 \ \cancel{torr}} \times 10^{-5} \ \cancel{torr}$ or 3.53×10^{14} molecules O_2.

V. CALCULATION OF MOLECULAR MASSES OF GASES

Example No. 6-21. 2.50 g of a gas occupy 500 mL at STP. What is the molecular mass of the gas?

The problem is to find the mass of 22.4 liters since this is the volume at STP occupied by 1 mole of the gas.

We have 2.50 g gas per 500 mL at STP or 2.50 g/500 mL. Changing mL to L we have

$$\frac{2.50 \ g}{500 \ \cancel{mL}} \times \frac{1000 \ \cancel{mL}}{1 \ L} \ or \ 5.00 \ g/L.$$

Then, changing L of gas to mol (1 mol of any gas at STP occupies 22.4 L),

$$5.00 \ g/\cancel{L} \times \frac{22.4 \ \cancel{L}}{1 \ mol} = 112 \ g/mol. \ Since \ 112 \ g \ of \ gas \ at \ STP \ occupy \ 22.4 \ L, \ the \ molecular \ mass \ is \ 112.$$

This may all be combined into one step, or

$$\frac{2.50 \ g}{500 \ \cancel{mL}} \times \frac{1000 \ \cancel{mL}}{1 \ \cancel{L}} \times \frac{22.4 \ \cancel{L}}{1 \ mol} = 112 \ g/mol.$$

Example No. 6-22.   2.40 g of a gas occupy 250 mL at 20°C and 740 torr. What is the molar mass of the gas?

We cannot use 22.4 L directly since this is the volume occupied by 1 mole of gas at STP and the gas does not meet these conditions. Therefore, the first step is to convert the volume to STP.

$$\text{Volume at STP} = 250 \text{ mL} \times \frac{273 \cancel{K}}{293 \cancel{K}} \times \frac{740 \cancel{torr}}{760 \cancel{torr}} = 227 \text{ mL}.$$

Thus, 2.40 g gas occupy 227 mL at STP. Then, changing mL to L to mol,

$$\frac{2.40 \text{ g}}{227 \cancel{mL}} \times \frac{1000 \cancel{mL}}{1 \cancel{L}} \times \frac{22.4 \cancel{L}}{1 \text{ mol}} = 237 \text{ g/mol, so that the molar mass is 237 g.}$$

Example No. 6-23.   If 7.50 g of a gas occupy 612 mL over water at 30°C and 745 torr, what is the molecular mass of the gas?

First we must convert the volume to STP using the vapor pressure of water at 30°C (31.8 mm Hg).

The partial pressure of the gas is 745.0 torr – 31.8 torr or 713.2 torr.

$$\text{Volume at STP} = 612 \text{ mL} \times \frac{713.2 \cancel{torr}}{760 \cancel{torr}} \times \frac{273 \cancel{K}}{303 \cancel{K}} = 518 \text{ mL}.$$

Therefore, 7.50 g gas occupies 518 mL at STP.

Changing mL to L to mol

$$\frac{7.50 \text{ g}}{518 \cancel{mL}} \times \frac{1000 \cancel{mL}}{1 \cancel{L}} \times \frac{22.4 \cancel{L}}{1 \text{ mol}} = 324 \text{ g/mol so that the molecular mass is 324.}$$

Example No. 6-24.   What will be the mass of 300 mL O_2 at STP?

We first change mL O_2 at STP to L O_2 to mol O_2, or

$$300 \cancel{mL} \times \frac{1 \cancel{L}}{1000 \cancel{mL}} \times \frac{1 \text{ mol}}{22.4 \cancel{L}} = 0.0134 \text{ mol } O_2.$$

Next, changing mol O_2 to g O_2 (1 mol O_2 has a mass of 32.0 g),

$$0.0134 \cancel{mol\,O_2} \times \frac{32.0 \text{ g } O_2}{1 \cancel{mol\,O_2}} = 0.429 \text{ g } O_2.$$

If this were all combined into one step,

$$300 \cancel{mL} \times \frac{1 \cancel{L}}{1000 \cancel{mL}} \times \frac{1 \cancel{mol}}{22.4 \cancel{L}} \times \frac{32.0 \text{ g}}{1 \cancel{mol}} = 0.429 \text{ g } O_2.$$

Example No. 6-25.   Calculate the density of N_2 at STP.

We know that 1 molar mass of N_2 at STP occupies 22.4 L so

$$28.0 \text{ g } N_2 \text{ occupies 22.4 L at STP}$$

$$D = \frac{M}{V} = \frac{28.0 \text{ g}}{22.4 \text{ L}} = 1.25 \text{ g/L at STP}$$

Example No. 6-26.   Calculate the mass of 10.0 liters of CH_4 at 27.0°C and 900 kPa.

We know that 1 mole of CH_4 or 16.0 grams occupy 22.4 liters at STP but the gas is not at standard conditions. Thus we have to convert the volume to STP before we can begin calculating.

$$\text{Volume at STP} = 10.0 \text{ liters} \times \frac{120 \text{ kPa}}{101.3 \text{ kPa}} \times \frac{273 \text{ K}}{300 \text{ K}} = 10.8 \text{ liters}.$$

Thus we have 16.0 g CH_4 per 22.4 L at STP, or $\dfrac{16.0 \text{ g } CH_4}{22.4 \text{ L}}$.

For 10.8 L at STP, we have $\dfrac{16.0 \text{ g } CH_4}{22.4 \text{ L}} \times 10.8 \text{ L} = 7.71 \text{ g } CH_4$.

Problem Assignment

6-1.   375 mL CO_2 at 20°C will occupy what volume at 40°C, pressure remaining constant?   (401 mL)

6-2.   350 mL of helium at 700 torr will occupy what volume at 900 mm Hg, temperature remaining constant? . . . . (272 mL)

6-3.   5.00 liters of O_2 at 30°C and 725 torr will occupy what volume at STP? . . . . (4.30 liters)

6-4.   200 mL of H_2 collected over water at 25°C and 716 torr will occupy what volume dry at STP? . . . . (167 mL)

6-5.   What are the relative rates of diffusion of methane, CH_4, and nitrogen, N_2 . . . . (1.32:1)

6-6.   0.750 g of a gas occupy 220 mL at STP. What is the molar mass of the gas? . . . . (76.4 g)

6-7.   Calculate the mass of 1.00 liter of carbon dioxide at STP. . . . . (1.96 g)

6-8.   A gas occupies a volume of 425 mL at 50°C and 100 kPa. What temperature will be required to change the volume to 300 mL, pressure remaining constant? . . . . (-45°C)

6-9.   A gas occupies a volume of 400 mL at 30°C and 726 torr. What pressure would be required to change the volume to 600 mL, temperature remaining constant? . . . . (484 torr)

6-10.   75.0 mL N_2 at 30°C and 820 torr over water will occupy what volume dry at the same temperature and pressure? . . . . (72.1 mL)

6-11.   If the temperature of 2.50 liters of NH_3 is changed from 20°C to 120°C and if the pressure is changed from 750 torr to 900 torr, what will be the new volume? . . . . (2.79 L)

6-12.   What is the molecular mass of a gas if 2.50 g of it occupy 1500 mL at 25°C and 729 torr? . . . . (42.5)

6-13.   Arrange the following gases in order of increasing rates of diffusion: Ar, CH_4, N_2, Ne, Cl_2, F_2, H_2, He, NH_3.

6-14.   200 mL of neon measured at STP will have what mass? . . . . (0.180 g)

6-15.   250 mL oxygen are collected over water at 27°C and 748 torr. What is the mass of this amount of oxygen? . . . . (0.308 g)

6-16.   How many molecules are present in 2.00 liters H_2 at 0°C and 1.00 torr? . . . . . . (7.07×10^{19})

6-17.   What are the relative rates of diffusion of chlorine and fluorine? . . . . (0.732:1)

6-18.   350 mL of a gas at 17°C and 795 torr become 450 mL at 35°C and what pressure? . . . . (657 torr)

6-19.   0.750 g of a gas collected over water at 29°C and 1.00 atm occupies 224 mL. What is the molecular mass of the gas? . . . . (86)

6-20. If 250 mL of H_2 diffuses through an orifice in a given period of time, how much O_2 will diffuse through the same size orifice in the same period of time? . . . . (62.5 mL)

6-21. 300 mL SO_3 at 0°C and 1.00 atm will occupy what volume at 100°C, pressure remaining constant?

6-22. 4.25 liters CO at −29°C and 755 torr will occupy what volume at 355 torr, temperature remaining constant?

6-23. 420 mL neon at 25°C and 815 torr will occupy what volume at STP?

6-24. 274 mL ethane gas at 27°C and 1.00 atmosphere pressure will occupy what volume at 227°C and 2.00 atmospheres pressure?

6-25. 200 mL H_2 collected over water at 18°C and 759 torr will occupy what volume dry at STP?

6-26. What will happen to the volume of a gas if we halve its pressure (from 1000 kPa to 500 kPa) and if we halve its Celsius temperature (from 100°C to 50°C)?

6-27. What volume will 60.0 g oxygen occupy at STP?

6-28. Calculate the mass of 252 liters SO_2 at STP.

6-29. Calculate the relative rates of diffusion of O_2 and NO_2.

6-30. What is the density of N_2 at STP? at 100°C and 700 torr?

6-31. A balloon holds 1.00×10^4 cubic feet of helium at 50°F and 1.00 atmosphere pressure. What will be the volume of the helium at 122°F and 0.750 atmospheres pressure?

6-32. A gas occupies 500 mL at 68.0°F and 720 torr. What will be its volume at 350K and 2.00 atmospheres pressure?

6-33. What is the molecular mass of a gas if 0.415 g of it occupy 180 mL at 15°C and 700 torr?

6-34. If 750 mL of oxygen escape per second through a small orifice, how rapidly will neon escape through the same size orifice?

6-35. What is the molecular mass of a gas if 0.750 g of it occupy 208 mL at 30°C and 800 torr?

6-36. 400 mL of air at 60°F and 18.0 lb./in.^{2} pressure will occupy what volume at STP? (1 atmosphere pressure = 14.7 lb./in.^{2}).

6-37. Gas A diffuses 2.50 times as fast as gas B. If the molecular mass of gas A is 30.0, what is the molecular mass of gas B?

6-38. If 400.00 mL CO are compressed to a volume of 198.90 mL when the pressure is increased from one atmosphere to two atmospheres, what is the percentage deviation from Boyle's Law?

6-39. 150 mL O_2 is collected over water at 17°C and 700 torr. If the dry volume becomes 100 mL and the pressure becomes 1.00 atmosphere, what will be the new temperature?

6-40. An automobile tire has a gauge pressure of 30.0 lb./in.^{2} (pressure in excess of atmospheric pressure) at 20°F. What will be the gauge pressure at 95°F?

6-41. A rigid balloon holds 5.000×10^4 cubic feet helium at 70°F and 1.00 atmosphere pressure. How much helium will escape if the balloon ascends to a point where the temperature falls to −10°F and the pressure to 10.0 torr?

6-42. Calculate the density of the following gases:
    a. H_2 at STP
    b. Ne at STP
    c. O_2 at STP
    d. O_2 at 0°C and 2.00 atmospheres pressure
    e. SO_2 at 100°C and 1.00 atmospheres pressure
    f. CH_4 at 100°C and 600 torr.

6-43. How many molecules are present in each of the following?
    a. 3.00 liters CO_2 at 0°C and 100 torr
    b. 200 mL N_2 at 0°C and 10.0 atmospheres pressure
    c. 100 mL H_2 at 0°C and 1.00×10^{-8} torr
    d. 2.00 liter O_2 at 273°C and 2.00 torr

6-44. How many atoms are present in:
    a. 6.50 g carbon
    b. 3.50 moles sulfur
    c. 1.85×10^{-4} g helium
    d. 8.27×10^{-5} moles zinc

6-45. How many molecules are present in:
    a. 35.0 g SO_2
    b. 22.4 g NH_3
    c. 3.00×10^{-15} moles H_2O
    d. 185 lb. Fe_3O_4
    e. 1.00 ounce $C_6H_{12}O_6$

6-46. A container holds 2.00 liters CH_4 at 20°C and 700 torr.
    a. How many moles does it contain?
    b. How many molecules does it contain?
    c. How many atoms does it contain?
    d. What is the mass of the gas?

6-47. Two liters of oxygen at 4 atmospheres pressure, four liters H_2 at 1 atmosphere pressure, and three liters N_2 at 5 atmospheres pressure are mixed and placed in a ten liter container. What will be the resulting pressure, assuming a constant temperature?

6-48. A container holds 1.50 g O_2 at STP.
    a. How many moles are present?
    b. How many molecules are present?
    c. How many atoms are present?
    d. What volume will the O_2 occupy?

6-49. A container holds 2.00 g CO_2 at 273°C and 1.00 atmosphere pressure.
    a. How many moles are present?
    b. How many molecules are present?
    c. How many atoms are present?
    d. What volume will the CO_2 occupy?

6-50. Helium is produced during the radioactive disintegration of radium. Each helium atom is formed from an alpha particle which in turn is counted by means of a Geiger counter. If the number of alpha particles is found to be 6.574×10^{14} and the volume of helium is 2.850×10^{-5} mL at 20°C and 700 torr, calculate Avogadro's number. (Use molar gas volume as 22.41 liters).

6-51. Given 22.4 liters of O_2 at STP.
   a. How many molecules are present?
   b. How many atoms are present?
   c. What volume is occupied by each molecule?
   d. What is the mass of each molecule?
   e. If the diameter of the oxygen molecule is 3.39×10^{-8} cm, what is the volume of an oxygen molecule?
   f. What percent of the container is empty space?

For the following problems, calculate using both the general gas law and van der Waals's equation.

	'a' $L^2 \cdot atm/mol^2$	'b' L/mol
He	0.0341	0.0237
H_2	0.244	0.0266
N_2	1.39	0.0394
O_2	1.36	0.0318
CCl_4	20.39	0.138
NH_3	4.17	0.0371

6-52. Calculate the pressure exerted by 4.00 moles of carbon dioxide at a temperature of 80.0°C and a volume of 3.00 liters. . . . . (38.6 atm, 33.5 atm)

6-53. What will be the pressure exerted by 12.00 g of helium at a temperature of 50.0 K if the volume is 30.0 liters?

6-54. Calculate the temperature at which 4.00 moles of hydrogen occupy a volume of 600 mL at a pressure of 100 atm. . . . . (183 K, 167 K)

6-55. Calculate the temperature at which 2.02 moles of carbon tetrachloride gas will occupy a volume of 100 liters at a pressure of 1.00 atm.

6-56. 0.250 moles of oxygen occupy a volume of 5.60 liters at a temperature of 0°C and a pressure of 1.00 atm. If the volume is decreased one hundredfold, what will be the new pressure?
. . . . (100 atm, 89.4 atm)

6-57. Calculate the pressure in Problem No. 6-56 if the volume is decreased five hundredfold.

6-58. 1.00 moles of ammonia occupy a volume of 22.4 liters at STP. What temperature will be required, holding volume constant, to increase the pressure tenfold? . . . . (2730 K, 2727 K)

6-59. Calculate the temperature in Problem No. 6-58 if the pressure is increased one hundredfold and the volume is decreased one hundredfold.

6-60. 2.00 moles of nitrogen at a temperature of 460 K occupy a volume of 0.400 liters at a pressure of 200 atm. Calculate the value of van der Waals 'b'. . . . . (0.0394)

6-61. Calculate the value of van der Waals' 'b' for a gas given the following data: volume 1.00 liter, pressure 50.0 atm, temperature 240 K, moles 3.00, and 'a' = 1.60 $L^2 \cdot atm/mol^2$.

Chapter 7

CALCULATION OF FORMULAS

The *empirical formula,* or the simplest formula of a compound, represents the relative number of each type of atom present in one molecule of the compound. The *molecular formula* represents the actual number of atoms of each type present in one molecule of the compound.

Thus, the empirical formula for acetylene is CH or 1 atom of Carbon for each atom of Hydrogen in the molecule. The molecular formula for acetylene is C_2H_2 which means that acetylene is composed of 2 atoms of Carbon and 2 atoms of Hydrogen in each molecule; its molecular mass is $2 \times 12.011 + 2 \times 1.008 = 26.038$.

The empirical formula does not represent the actual number of atoms present and so cannot represent the molecular mass. The molecular formula is always a simple integral multiple (1, 2, 3, etc.) of the empirical formula.

Example No. 7-1. Determine the empirical formula of a compound containing 11.19% Hydrogen and 88.81% Oxygen.

For every 100 grams of compound, there is 11.19% or 11.19 g Hydrogen and 88.81% or 88.81 g Oxygen. If we divide each mass by the corresponding atomic mass (to four significant figures), we will have the number of moles of each present in the compound.

For Hydrogen, $11.19 \text{ g} \times \dfrac{1 \text{ mol}}{1.008 \text{ g}} = 11.10$ mol of H atoms

For Oxygen, $88.81 \text{ g} \times \dfrac{1 \text{ mol}}{16.00 \text{ g}} = 5.550$ mol of O atoms

So we have 11.10 moles Hydrogen to 5.550 moles Oxygen. Dividing by the smaller number of moles (to arrive at a whole number) we have

For Hydrogen, $\dfrac{11.10}{5.550}$ mol = 2 mol

For Oxygen, $\dfrac{5.550}{5.550}$ mol = 1 mol

or a ratio of 2 moles Hydrogen to 1 mole Oxygen. Thus, the empirical formula is H_2O or OH_2. It is usually best to write the formula listing the elements in the order given in the problem, so the empirical formula is H_2O.

NOTE: We used the actual atomic masses in our calculations. Our results would have been exactly the same if we had used the atomic masses which had been rounded off to whole numbers. Thus, for Hydrogen we could have used the atomic mass 1 and for Oxygen the atomic mass 16.

Example No. 7-2. Calculate the empirical formula of a compound containing 43.64% Phosphorous and 56.36% Oxygen.

Here we will use the rounded-off atomic masses,

Phosphorus, P=31 and Oxygen, O=16.

In 100 grams of the compound, we will have:

$$\text{For Phosphorus, } 43.64 \text{ g} \times \frac{1 \text{ mol}}{31 \text{ g}} = 1.4 \text{ mol}$$

$$\text{For Oxygen, } 56.36 \text{ g} \times \frac{1 \text{ mol}}{16 \text{ g}} = 3.5 \text{ mol}$$

Dividing by the smaller number of moles,

$$\text{For Phosphorus, } \frac{1.4}{1.4} \text{ mol} = 1 \text{ mol}$$

$$\text{For Oxygen, } \frac{3.5}{1.4} \text{ mol} = 2.5 \text{ mol}$$

Since we can never have fractional values, we must double our results to get whole numbers. Thus we have:

$$\text{For Phosphorus, } 2 \times 1 \text{ mol} = 2 \text{ mol}$$

$$\text{For Oxygen, } 2 \times 2.5 \text{ mol} = 5 \text{ mol}$$

giving the empirical formula P_2O_5.

Example No. 7-3.   A certain compound contains 26.5% Potassium, 35.4% Chromium, and 38.1% Oxygen. What is its empirical formula?

In each 100 grams, we have:

$$\text{For Potassium, K, } 26.5 \text{ g} \times \frac{1 \text{ mol}}{39 \text{ g}} = 0.68 \text{ mol}$$

$$\text{For Chromium, Cr, } 35.4 \text{ g} \times \frac{1 \text{ mol}}{52 \text{ g}} = 0.68 \text{ mol}$$

$$\text{For Oxygen, O, } 38.1 \text{ g} \times \frac{1 \text{ mol}}{16 \text{ g}} = 2.38 \text{ mol}$$

Dividing by the smallest number of moles (0.68), we have

$$\text{For Potassium, K, } \frac{0.68}{0.68} \text{ mol} = 1 \text{ mol}$$

$$\text{For Chromium, Cr, } \frac{0.68}{0.68} \text{ mol} = 1 \text{ mol}$$

$$\text{For Oxygen, O, } \frac{2.38}{0.68} \text{ mol} = 3.5 \text{ mol}$$

Since we cannot have fractions, we double all the results and get

$$\text{For K, } 2 \times 1 \text{ mol} = 2 \text{ mol}$$

$$\text{For Cr, } 2 \times 1 \text{ mol} = 2 \text{ mol}$$

$$\text{For O, } 2 \times 3.5 \text{ mol} = 7 \text{ mol}$$

Thus, the empirical formula is $K_2Cr_2O_7$.

Example No. 7-4. A compound of molecular mass 56.108 has the following composition: Carbon 85.63% and Hydrogen 14.37%. What is its molecular formula?

First let us calculate the empirical formula of the compound.

$$\text{For Carbon, C,} \quad 85.63\,\cancel{g} \times \frac{1\text{ mol}}{12\,\cancel{g}} = 7.14\text{ mol}$$

$$\text{For Hydrogen, H,} \quad 14.37\,\cancel{g} \times \frac{1\text{ mol}}{1.0\,\cancel{g}} = 14.4\text{ mol}$$

and dividing by the smaller number of moles we get

$$\text{For C,} \quad \frac{7.14}{7.14}\text{ mol} = 1\text{ mol}$$

$$\text{For H,} \quad \frac{14.4}{7.14}\text{ mol} = 2\text{ mol}$$

so the empirical formula is CH_2.

The mass of this simplest formula, CH_2 is

$$1 \times 12.011 + 2 \times 1.008 = 14.027.$$

The molecular mass of the compound is 56.108 and, therefore, it must contain 56.108/14.027 or 4 times the amount given by the empirical formula. Thus the molecular formula is

$$C_4H_8$$

For a check, $4 \times 12.011 + 8 \times 1.008$ does equal 56.108.

Example No. 7-5. 0.4655 grams of iron when combined with sulfur yielded 1.0000 grams iron sulfide. What is the empirical formula of this substance?

Since 0.4655 grams of iron gave 1.0000 grams of sulfide, there must have been 1.0000 − 0.4655 or 0.5345 grams of sulfur present. Then,

$$\text{For Iron, Fe,} \quad 0.4655\,\cancel{g} \times \frac{1\text{ mol}}{55.8\,\cancel{g}} = 0.00834\text{ mol}$$

$$\text{For Sulfur, S,} \quad 0.5345\,\cancel{g} \times \frac{1\text{ mol}}{32\,\cancel{g}} = 0.017\text{ mol}$$

and dividing by the smaller, we have

$$\text{For Fe,} \quad \frac{0.00834}{0.00834}\text{ mol} = 1\text{ mol, and}$$

$$\text{For S,} \quad \frac{0.017}{0.00834}\text{ mol} = 2\text{ mol}$$

so the simplest formula is FeS_2.

Example No. 7-6. 5.00 grams of a hydrated salt $MgSO_4 \cdot xH_2O$ lost 2.56 grams of water of crystallization upon heating. What is the formula of this hydrate?

This is the same type of problem as on the previous page, except that here we are working with molecules instead of atoms.

There must be 5.00 g – 2.56 g or 2.44 g anhydrous $MgSO_4$ left after the water is driven off, so the compound contains

$$2.44 \text{ grams } MgSO_4 \text{ and } 2.56 \text{ grams } H_2O.$$

1 mole $MgSO_4$ weighs 120 g; 1 mole H_2O weighs 18.0 g. Then,

$$\text{For } MgSO_4 \text{ we have } 2.44 \text{ g} \times \frac{1 \text{ mol}}{120 \text{ g}} = 0.0203 \text{ mol}$$

$$\text{For } H_2O \quad \text{we have } 2.56 \text{ g} \times \frac{1 \text{ mol}}{18.0 \text{ g}} = 0.142 \text{ mol}$$

and dividing by the smaller number of mol, we have

$$\text{For } MgSO_4, \quad \frac{0.0203 \text{ mol}}{0.0203} = 1 \text{ mol}$$

$$\text{For } H_2O, \quad \frac{0.142 \text{ mol}}{0.0203} = 7 \text{ mol}$$

so the formula of the hydrate must be $MgSO_4 \cdot 7H_2O$.

Example No. 7-7.   1.000 liter of an oxide of Nitrogen at STP weighs 4.107 g and contains 30.43% Nitrogen, the rest being Oxygen. What is the molecular formula of the oxide?

First let us calculate the empirical formula.

$$\text{For N, we have} \quad 30.43 \text{ g} \times \frac{1 \text{ mol}}{14.0 \text{ g}} = 2.17 \text{ mol}$$

$$\text{For O, we have } (100 - 30.43) \text{ g} \times \frac{1 \text{ mol}}{16.0 \text{ g}} = 4.34 \text{ mol}$$

(On the basis that we have 100 grams of sample, 30.43% or 30.43 grams is Nitrogen and the rest or 100 – 30.43 is Oxygen.)

and dividing by the smaller, we have

$$\text{For N, } \frac{2.17}{2.17} \text{ mol} = 1 \text{ mol}$$

$$\text{For O, } \frac{4.34}{2.17} \text{ mol} = 2 \text{ mol}$$

so the empirical formula is NO_2.

From the data, 1 liter weighs 4.107 grams at STP so 22.4 liters will weigh $22.4 \times 4.107 \text{ g} = 92.0 \text{ g}$ at STP. This is the molar mass of the compound since it is the mass of 22.4 liters of the gas at STP. The mass of the atoms in the empirical formula is.

$$(NO_2) = 1 \times 14.0 + 2 \times 16.0 = 46.0 \text{ g}$$

Therefore, there must be $\frac{92.0}{46.0}$ or 2 of the simplest units in the molecular formula which thus is N_2O_4.

7-54

Problem Assignment

7-1. A certain compound contains 49.5% manganese and 50.5% oxygen. What is its empirical formula? . . . . (Mn_2O_7)

7-2. 2.00 g sulfur combine with 2.00 g oxygen. What is the simplest formula for the oxide produced? . . . . (SO_2)

7-3. Sodium tetrathionate contains 17.0% sodium, 47.4% sulfur, and 35.6% oxygen. If its molecular mass is 270, what is its molecular formula? . . . . $(Na_2S_4O_6)$

7-4. 2.84 grams of S candium when burned in air yield 4.36 g oxide. What is the empirical formula for this oxide?

7-5. A compound of molecular mass 198 contains 64% copper and 36% chlorine. What is its molecular formula?

7-6. 25.00 g of a hydrate of $CuSO_4$ lost 9.02 g on heating. What is the formula of the hydrate? . . . . $(CuSO_4 \cdot 5H_2O)$

7-7. 2.00 liters of a gas at STP weigh 5.00 g. The gas contains 85.7% carbon and 14.3% hydrogen. What is the molecular formula of the gas? . . . . (C_4H_8)

7-8. 80.000 g of a hydrate lost 14.698 g on heating. The anhydrous salt contains 21.800 g calcium, 0.548 g hydrogen, 16.847 g phosphorus, the rest being oxygen. What is the simplest formula of the hydrate?

7-9. 200 mL of a gas at 30°C and 755 torr weigh 416 mg. The gas contains 92.26% carbon and 7.74% hydrogen. What is its molecular formula?

7-10. 4.00 g of sodium reacted with sulfur to yield 6.78 g of a compound of molecular mass 78. What is the molecular formula of the sulfide?

What are the empirical formulas for the following chlorides?

7-11. A chloride containing 31.1% sulfur . . . . (SCl_2)

7-12. A chloride containing 14.9% phosphorus . . . . (PCl_5)

7-13. A chloride containing 7.79% carbon . . . . (CCl_4)

7-14. A chloride containing 34.3% iron

7-15. A chloride containing 29.7% arsenic

What are the molecular formulas for the following compounds?

Molecular mass

7-16. 64.2% Cu and 35.8% Cl. . . . . 198.1 . . . . (Cu_2Cl_2)

7-17. 78.3% B and 21.7% H . . . . 27.6 . . . . (B_2H_6)

7-18. 43.7% P and 56.3% O . . . . 284 . . . . (P_4O_{10})

7-19. 21.85% Mg, 27.83% P, 50.32% O 222.6

7-20. 10.7% Na, 59.3% I, 29.9% O 214

7-21. 36.09% C, 5.30% H, 48.08% O, 10.53% N . . . . 133.1 . . . . $(C_4H_7O_4N)$

7-22. 40.81% C, 6.17% H, 43.50% O, 9.52% N . . . . 147.1 . . . . $(C_5H_9O_4N)$

7-23. 40.44% C, 7.92% H, 35.92% O, 15.72% N 189.1

7-24. 49.29% C, 9.65% H, 21.89% O, 19.17% N 146.2

7-25. 23.19% C, 1.43% H, 8.24% O, 1.80% N, 65.34% I 776.8

CALCULATION OF PERCENT COMPOSITION FROM MOLECULAR FORMULAS

If a compound contains 2 grams of Sulfur and 3 grams of Oxygen, there are 2 grams of Sulfur out of a total of 5 grams of compound and the fraction that is Sulfur is 2/5 of the total. Thus 2/5 or 0.4 of every sample of the compound is Sulfur. If we take 1 gram of the compound, there are 0.4 × 1 gram or 0.4 grams of Sulfur present; if we take 100 grams of the compound, there are 0.4 × 100 grams or 40 grams Sulfur present. Also, the fraction that is Oxygen is 3/5 of the total or 0.6. The sum of these fractions must equal the whole or 1; 2/5 + 3/5 = 5/5 = 1.

Percent represents the parts per hundred, so, therefore, it is 100 times the fraction of the substance present per 1 unit of weight. Thus,

the percent Sulfur is 2/5 × 100 or 40%, and
the percent Oxygen is 3/5 × 100 or 60%.

The sum in this case must equal the whole or 100%.

NOTE: 2/5, 0.4, 40/100, 40% all have the same meaning.

Example No. 7-8. Calculate the percent composition of $KClO_3$.

First let us calculate the molecular mass of the compound (to one decimal place)

$$
\begin{array}{rcl}
1 \ K & = & 39.1 \\
1 \ Cl & = & 35.5 \\
3 \ O & = & 48.0 \\
\hline
\text{molecular mass } KClO_3 & = & 122.6
\end{array}
$$

$$\text{Percent K} = \frac{\text{amount of K present}}{\text{mass of compound}} \times 100\% = \frac{1 \text{ molecular mass K}}{\text{molecular mass compound}} \times 100\% = \frac{39.1}{122.6} \times 100\% = 31.9\%$$

$$\text{Percent Cl} = \frac{1 \text{ molecular mass Cl}}{\text{molecular mass compound}} \times 100\% = \frac{35.5}{122.6} \times 100\% = 28.9\%$$

$$\text{Percent O} = \frac{3 \text{ molecular masses O}}{\text{molecular mass compound}} \times 100\% = \frac{48.0}{122.6} \times 100\% = 39.2\%$$

Total percent = 31.9% + 28.9% + 39.2% = 100.0%

We should note here that the total percent may be close to but not equal to 100%. This may be due to not obtaining enough significant figures or not rounding off properly.

Example No. 7-9. Determine the percent S in SO_2 and in SO_3.

Molecular mass SO_2 = 32.1 + 32.0 = 64.1

Molecular mass SO_3 = 32.1 + 48.0 = 80.1

Percent S in SO_2 = $\dfrac{32.1}{64.1} \times 100\% = 50.1\%$

Percent S in SO_3 = $\dfrac{32.1}{80.1} \times 100\% = 40.1\%$

Example No. 7-10. How many kilograms of mercury are contained in a 5.000 kilogram sample of HgO?

First, let us calculate the percent of Mercury in HgO and from this figure we can get the amount of Hg present in the sample.

Molecular mass HgO = 200.6 + 16.0 = 216.6

$$\text{Percent Hg} = \frac{200.6}{216.6} \times 100\% = 92.61\%$$

Therefore, the amount of mercury present in a 5.000 kg sample is

92.61% of 5.000 kg or $\frac{92.61}{100} \times 5.000$ kg = 4.630 kg.

Example No. 7-11. Calculate the percent composition of $Na_2SO_4 \cdot 10H_2O$ (calculate the percent of water as a unit).

Molecular mass of compound = $2 \times 23.0 + 32.1 + 4 \times 16.0 + 10 \times 18.0 = 322.1$

Percent Na	$= \dfrac{2 \times 23.0}{322.1} \times 100\%$	$= 14.3\%$
Percent S	$= \dfrac{32.1}{322.1} \times 100\%$	$= 10.0\%$
Percent O	$= \dfrac{4 \times 16.0}{322.1} \times 100\%$	$= 19.8\%$ (excluding water)
Percent H_2O	$= \dfrac{10 \times 18.0}{322.1} \times 100\%$	$= 55.9\%$
Total percent		$= 100.0\%$

Example No. 7-12. What mass of Aluminum is contained in 1.0 ton of bauxite which is 87% Al_2O_3?

molecular mass $Al_2O_3 = 2 \times 26.98 + 3 \times 16.00 = 101.96$

$$\text{Percent Aluminum} = \frac{2 \times 26.98}{101.96} \times 100\% = 52.92\% \text{ aluminum in } Al_2O_3$$

Since the ore is only 87% pure, there is only 87% of 1 ton or 0.87 tons Al_2O_3 present. The mass of Aluminum in this ore is

52.92% × 0.87 tons = 0.5292 × 0.87 tons = 0.46 tons.

Example No. 7-13. Calculate the percent composition of an alloy containing 13.25 lb. Bi, 8.00 lb. Pb and 3.75 lb. Sn.

First, let us find the total mass of the alloy.

13.25 lb. + 8.00 lb. + 3.75 lb. = 25.00 lb.

Percent Bi	$= \dfrac{\text{mass Bi}}{\text{mass of alloy}} \times 100\%$	$= \dfrac{13.25 \text{ lb}}{25.00 \text{ lb}} \times 100\%$	$= 53.0\%$
Percent Pb	$= \dfrac{\text{mass Pb}}{\text{mass of alloy}} \times 100\%$	$= \dfrac{8.00 \text{ lb}}{25.00 \text{ lb}} \times 100\%$	$= 32.0\%$
Percent Sn	$= \dfrac{\text{mass Sn}}{\text{mass of alloy}} \times 100\%$	$= \dfrac{3.75 \text{ lb}}{25.00 \text{ lb}} \times 100\%$	$= 15.0\%$
Total percent			$= 100.0\%$

Problem Assignment

Calculate the percent composition of the following compounds (to 1 decimal place):

7-26. NH_3; PH_3 .... (82.4% N, 17.6% H; 91.2% P, 8.8% H)

7-27. Fe_3O_4; SnO .... (72.3% Fe, 27.3% O; 88.1% Sn, 11.9% H)

7-28. H_3PO_4 .... (3.1% H, 31.6% P, 65.3% O)

7-29. $MgSO_4$; $C_{12}H_{22}O_{11}$

7-30. K_2PtCl_6; $C_6H_{10}O_5$

Calculate the % K in problems 7-31 through 7-36.

7-31. K_2S .... (70.9%)

7-32. K_2SO_3 .... (49.4%)

7-33. K_2SO_4 .... (44.9%)

7-34. $K_2S_2O_7$

7-35. $KHSO_4$

7-36. $K_3Fe(CN)_6$

7-37. What is the percent water in $Na_2S_4O_6 \cdot 2H_2O$? .... (11.8%)

7-38. What is the total percent Oxygen in $Na_2SO_4 \cdot 10H_2O$ .... (69.6%)

7-39. What is the percent nitrogen in NH_4NO_2? .... (43.8%)

7-40. What is the percent carbon in $C_6H_{12}O_6$; in C_6H_6? .... (40.0%, 92.3%)

7-41. What is the percentage composition of an alloy containing 23.20 lb. lead, 10.40 lb. tin, 6.00 lb. antimony, and 0.40 lb. copper?

7-42. What is the percent water in $NaAl(SO_4)_2 \cdot 12H_2O$?

7-43. Which contains a higher percentage of chlorine, KClO or $NaClO_2$?

7-44. What is the percent tungsten in $Na_6W_7O_{24} \cdot 16H_2O$?

7-45. What is the percent Cr in: $CrCl_3$, K_2CrO_4, $Na_2Cr_2O_7$, CrO_3, Cr_2O_3?

Chapter 8

MASS AND VOLUME RELATIONS IN CHEMICAL CALCULATIONS

Stoichiometry is that branch of chemistry which deals with the relationships involved between substances in chemical reactions. A balanced equation is absolutely necessary because it represents the quantitative facts of experiment.

1. It tells what the reactants are and what the products are.

2. It tells how many grams of each of the reactants are required to give the same total number of grams of products (Law of Conservation of Mass).

3. It tells in what ratio the reactants combine and also in what ratio the products are formed (Law of Definite Proportions).

4. It tells the relative number of atoms and molecules reacting.

5. It tells the volumes (in the case of gases) of the reactants and the products.

In the equation $\qquad$ $2KClO_3 \qquad \rightarrow \qquad 2KCl + 3O_{2(g)}$

we see that

2 molecules	$\rightarrow$	2 molecules + 3 molecules
2 molecular masses	$\rightarrow$	2 molecular masses + 3 molecular masses
2×122.6 mass units	$\rightarrow$	2×74.6 mass units + 3×32 mass units
2 molar masses	$\rightarrow$	2 molar masses + 3 molar masses
2×122.6 g	$\rightarrow$	2×74.6 g + 3×32 g
2 molar masses	$\rightarrow$	2 molar masses + 3 molar volumes at STP
2×122.6 g	$\rightarrow$	2×74.6 g + 3×22.4 liters at STP
2 mol	$\rightarrow$	
2 mol + 3 mol		

The last relationship, the molar one, is the one best suited to stoichiometrical calculations because it deals with whole numbers of moles as indicated by the coefficients in the original equation. Let us see how to use this relationship.

Example No.8-1.   From the balanced equation.

$3 Cu + 8 HNO_3 \rightarrow 3 Cu(NO_3)_2 + 2NO_{(g)} + 4 H_2O$

Calculate:

a)   the number of moles of $Cu(NO_3)_2$ produced from 5.00 moles of HNO_3

b)   the number of moles of NO produced from 5.00 moles of HNO_3

c)   the number of grams of Cu to react with 5.00 moles of HNO_3

d)   the number of liters of NO at STP produced from 5.00 moles of HNO_3

e)   the number of grams of NO produced from 2.00 moles of Cu

a) The equation tells us that 8 moles HNO_3 yield 3 moles $Cu(NO_3)_2$, so using this as a conversion factor for moles of HNO_3 to moles $Cu(NO_3)_2$

$$5.00 \; \text{mol HNO}_3 \times \frac{3 \text{ mol Cu(NO}_3)_2}{8 \text{ mol HNO}_3} = 1.88 \text{ mol Cu(NO}_3)_2$$

Note that the unwanted units cancel.

b) Next, since 8 moles HNO_3 yield 2 moles NO,

$$5.00 \; \text{mol HNO}_3 \times \frac{2 \text{ mol NO}}{8 \text{ mol HNO}_3} = 1.25 \text{ mol NO}$$

c) Then, 8 moles HNO_3 react with 3 moles Cu, so

$$5.00 \; \text{mol HNO}_3 \times \frac{3 \text{ mol Cu}}{8 \text{ mol HNO}_3} = 1.88 \text{ mol Cu}$$

Next, since 1 mole Cu has a mass of 63.5 g

$$1.88 \; \text{mol Cu} \times \frac{63.5 \text{ g Cu}}{1 \text{ mol Cu}} = 119 \text{ g Cu}$$

Note that all of the conversions could have been included in one equation:

$$5.00 \; \text{mol HNO}_3 \times \frac{3 \text{ mol Cu}}{8 \text{ mol HNO}_3} \times \frac{63.5 \text{ g Cu}}{1 \text{ mol Cu}} = 119 \text{ g Cu}$$

d) From part b above we see that 5.00 moles HNO_3 yield 1.25 moles NO. (Since NO is a gas, 1 mole of it will occupy 22.4 liters at STP.) Thus,

$$1.25 \; \text{mol NO} \times \frac{22.4 \text{ L}}{1 \text{ mol}} = 28.0 \text{ L NO at STP.}$$

e) We can change moles of Cu to moles of NO to grams of NO by using the conversions $\frac{2 \text{ mol NO}}{3 \text{ mol Cu}}$ and $\frac{30.0 \text{ g NO}}{1 \text{ mol NO}}$ respectively.

Therefore, $2.00 \; \text{mol Cu} \times \dfrac{2 \text{ mol NO}}{3 \text{ mol Cu}} \times \dfrac{30.0 \text{ g NO}}{1 \text{ mol NO}} = 40.0 \text{ g NO}$

Example No. 8-2. What mass of MgO will be produced by burning 1.75 g Mg? The balanced equation is: $2 \text{ Mg} + O_2 \rightarrow 2 \text{ MgO}$

First we have to change grams Mg to moles Mg (1 mole Mg has a mass of 24.3 g). Then we change moles Mg to moles MgO (the equation tells us that 2 moles Mg produce 2 moles MgO). Finally we must change moles MgO to grams MgO (1 mole MgO has a mass of 40.3 g). Combining all of this information we have:

$$1.75 \; \text{g Mg} \times \frac{1 \text{ mol Mg}}{24.3 \text{ g Mg}} \times \frac{2 \text{ mol MgO}}{2 \text{ mol Mg}} \times \frac{40.3 \text{ g MgO}}{1 \text{ mol MgO}} = 2.90 \text{ g MgO.}$$

$$\text{g Mg} \rightarrow \quad \text{mol Mg} \rightarrow \quad \text{mol MgO} \rightarrow \quad \text{g MgO}$$

Note that if any of the conversion factors had been inverted, the units would not have cancelled indicating an error in the setup of the problem.

Example No. 8-3.   How many grams Fe_3O_4 will be produced by passing 200 grams steam over hot iron?

First we write a balanced equation:

$$3\ Fe + 4\ H_2O \rightarrow Fe_3O_4 + 4\ H_2$$

Since the equation indicates relationships in terms of moles, we should work in those units. Therefore, we first change grams H_2O to moles H_2O (1 mol H_2O has a mass of 18.0 g). Then we change moles H_2O to moles Fe_3O_4 (from the balanced equation we see that 4 moles H_2O yield 1 mole Fe_3O_4). Finally, we change moles Fe_3O_4 to grams Fe_3O_4 (1 mole Fe_3O_4 has a mass of 232 g).

Thus we have:

$$200\ \text{g } H_2O \times \frac{1\ \text{mol } H_2O}{18.0\ \text{g } H_2O} \times \frac{1\ \text{mol } Fe_3O_4}{4\ \text{mol } H_2O} \times \frac{232\ \text{g } Fe_3O_4}{1\ \text{mol } Fe_3O_4} = 644\ \text{g } Fe_3O_4$$

Example No. 8-4.   How many grams $KClO_3$ will have to be heated to produce 5.60 grams O_2? How much KCl will be produced?

The balanced equation is: $2\ KClO_3 \rightarrow 2\ KCl + 3\ O_2$

First we change grams O_2 to moles O_2 (1 mole O_2 has a mass of 32.0 g). Then we change moles O_2 to moles $KClO_3$ (2 moles $KClO_3$ yield 3 moles O_2). Finally we change moles $KClO_3$ to grams $KClO_3$ (1 mole $KClO_3$ has a mass of 123 g).

Thus we have:

$$5.60\ \text{g } O_2 \times \frac{1\ \text{mol } O_2}{32.0\ \text{g } O_2} \times \frac{2\ \text{mol } KClO_3}{3\ \text{mol } O_2} \times \frac{123\ \text{g } KClO_3}{1\ \text{mol } KClO_3} = 14.3\ \text{g } KClO_3$$

Next, to see how much KCl is produced, we convert grams of O_2 to moles O_2 (1 mole of O_2 has a mass of 32.0 g) to moles KCl (3 moles O_2 are produced along with 2 moles KCl) to grams KCl (1 mole KCl has a mass of 74.6 g).

Thus we have:

$$5.60\ \text{g } O_2 \times \frac{1\ \text{mol } O_2}{32.0\ \text{g } O_2} \times \frac{2\ \text{mol } KCl}{3\ \text{mol } O_2} \times \frac{73.6\ \text{g } KCl}{1\ \text{mol } KCl} = 8.70\ \text{g } KCl$$

Example No. 8-5.   How many liters H_2 at STP will be produced by the action of 50.0 g Na upon water?

$$2\ Na + 2\ H_2O \rightarrow 2\ NaOH + H_{2(g)}$$

First we change grams Na to moles Na (1 mol Na has a mass of 23.0 g). Then we change moles Na to moles H_2 (2 moles Na produce 1 mole H_2). Finally we change moles H_2 to liters H_2 at STP (1 mole H_2 at STP occupies 22.4 L).

Thus we have:

$$50.0\ \text{g Na} \times \frac{1\ \text{mol Na}}{23.0\ \text{g Na}} \times \frac{1\ \text{mol } H_2}{2\ \text{mol Na}} \times \frac{22.4\ \text{L } H_2\ \text{STP}}{1\ \text{mol } H_2} = 24.3\ \text{L } H_2\ \text{at STP.}$$

Example No. 8-6.   How many grams NaCl must be electrolyzed to give 200 L Cl_2 at STP?

The balanced equation is:

$$2\ NaCl \xrightarrow{\text{electric current}} 2\ Na + Cl_2$$

We can change liters Cl_2 at STP to moles Cl_2 (1 mol Cl_2 at STP occupies 22.4 L). Then, we can change moles Cl_2 to moles NaCl (2 moles NaCl yield 1 mole Cl_2) and finally to grams NaCl (1 mole NaCl has a mass of 58.5 g).

So we have:

$$200 \text{ L } Cl_2 \text{ STP} \times \frac{1 \text{ mol } Cl_2}{22.4 \text{ L } Cl_2 \text{ STP}} \times \frac{2 \text{ mol NaCl}}{1 \text{ mol } Cl_2} \times \frac{58.5 \text{ g NaCl}}{1 \text{ mol NaCl}} = 1.04 \times 10^3 \text{ g NaCl}$$

$$\text{L } Cl_2\text{STP} \quad \rightarrow \quad \text{moles } Cl_2\text{STP} \quad \rightarrow \quad \text{moles NaCl} \quad \rightarrow \quad \text{g NaCl}$$

Example No. 8-7. How many liters NH_3 will be produced by the reaction of 20 liters of N_2 with excess H_2?

The balanced equation for the reaction is:

$$N_{2(g)} + 3 H_{2(g)} \rightarrow 2 NH_{3(g)}$$

Let us assume that all the gases are at STP. Then we can change liters N_2 to moles N_2 (1 mol N_2 at STP occupies 22.4 L) to moles NH_3 (1 mole N_2 yields 2 moles NH_3) to liters NH_3 (1 mole NH_3 at STP occupies 22.4 L). Or,

$$20 \text{ L } N_2 \times \frac{1 \text{ mol } N_2}{22.4 \text{ L } N_2} \times \frac{2 \text{ mol } NH_3}{1 \text{ mol } N_2} \times \frac{22.4 \text{ L } NH_3}{1 \text{ mol } NH_3} = 40 \text{ L } NH_3$$

If the gases had not been at STP, then instead of 22.4 L in both cases we would have had some other number. However, this other number (as well as the 22.4 in the above example) will cancel anyway. Thus the volume of NH_3 produced from 20 L N_2 will still be 40 L.

However, let us consider the reaction again. It states that:

$$N_2 + 3 H_2 \quad \rightarrow 2 NH_3$$

$$1 \text{ mol} + 3 \text{ mol} \quad \rightarrow 2 \text{ mol and since they are all gases,}$$

$$22.4 \text{ L} + 3 \times 22.4 \text{ L} \rightarrow 2 \times 22.4 \text{ L and dividing by } 22.4 \text{ L,}$$

$$1 \text{ volume} + 3 \text{ volumes} \quad \rightarrow 2 \text{ volumes}$$

Thus 1 volume of N_2 will react with 3 volumes of H_2 to give 2 volumes of NH_3 since they are all gases. Note that these numbers are the same as the coefficients in the balanced equation. Thus,

$$20 \text{ L } N_2 \times \frac{2 \text{ volumes } NH_3}{1 \text{ volume } N_2} = 40 \text{ L } NH_3 \text{ as was calculated above.}$$

Example No. 8-8. In the reaction $2 C_2H_2 + 5 O_2 \rightarrow 4 CO_2 + 2 H_2O$ (all gases)

(a) what volume of CO_2 will be produced by burning 50.0 L C_2H_2?
(b) what volume of O_2 will be needed?

Since they are all gases, we can use the volumes directly from the coefficients in the balanced equation.

(a) The equation states that 2 volumes C_2H_2 yield 4 volumes CO_2, so

$$50.0 \text{ L } C_2H_2 \times \frac{4 \text{ volumes } CO_2}{2 \text{ volumes } C_2H_2} = 100 \text{ L } CO_2$$

(b) The equation also states that 2 volumes C_2H_2 require 5 volumes O_2, so

$$50.0 \text{ L } C_2H_2 \times \frac{5 \text{ volumes } O_2}{2 \text{ volumes } C_2H_2} = 125 \text{ L } O_2$$

Note that if we had changed the liters C_2H_2 to mole C_2H_2 to mole O_2 to liters O_2 the answer would have been the same although the previous method is much simpler.

(assuming STP) $50.0 \; \cancel{L \; C_2H_2} \times \dfrac{1 \; \cancel{mol \; C_2H_2}}{22.4 \; \cancel{L \; C_2H_2}} \times \dfrac{5 \; \cancel{mol \; O_2}}{2 \; \cancel{mol \; C_2H_2}} \times \dfrac{22.4 \; L \; O_2}{1 \; \cancel{mol \; O_2}} = 125 \; L \; O_2$

We must be very careful in problems of this type that the substances we are dealing with are gases. If one is a solid, we must use the methods outlined in examples 8-1 to 8-6.

In many reactions, an excess of one reagent is usually supplied to make sure that the process goes to completion. However, sometimes we do not know which of the reactants is in excess and so have to calculate which is the determining factor before we can begin our work.

Example No. 8-9. If 4.00 grams aluminum are reacted with 10.0 grams ferric oxide, how many grams of iron will be produced?

First, $\qquad\qquad 2 \; Al + Fe_2O_3 \rightarrow Al_2O_3 + 2 \; Fe.$

Now we have to decide whether to base our calculations upon the fact that 2 moles Al yield 2 moles Fe or upon the fact that 1 mole Fe_2O_3 yields 2 moles Fe. Obviously, unless both reactants are present in exactly the proper amount, one of these must be in excess and so we cannot use it in our calculations.

Let us calculate the number of moles of each reactant.

$4.00 \; g \; Al \times \dfrac{1 \; mol}{27.0 \; g} = 0.148 \; moles \; Al \qquad\qquad 10.0 \; g \; Fe_2O_3 \times \dfrac{1 \; mol}{160 \; g} = 0.0625 \; mol \; Fe_2O_3$

Next let us calculate how many moles of each are required to react with the given amount of the other.

for the 0.148 mol Al $\qquad\qquad\qquad\qquad$ for the 0.0625 mol Fe_2O_3

$0.148 \; \cancel{mol \; Al} \times \dfrac{1 \; mol \; Fe_2O_3}{2 \; \cancel{mol \; Al}} = 0.0740 \; mol \; Fe_2O_3 \qquad 0.0625 \; \cancel{mol \; Fe_2O_3} \times \dfrac{2 \; mol \; Al}{1 \; \cancel{mol \; Fe_2O_3}} = 0.125 \; mol \; Al$

Thus, 0.148 mol Al require 0.0740 mol Fe_2O_3 and 0.0625 mol Fe_2O_3 require 0.125 mol Al. If we use the 0.148 mol Al we need 0.0740 mol Fe_2O_3 and we see that we don't have that much. That is, there is an excess of Al present. If we use the 0.0625 mol Fe_2O_3 then we need 0.125 mol Al which we do have (again we see that we have an excess of Al present). Therefore, the Fe_2O_3 is the limiting factor (once all of it has reacted, no further reaction takes place).

Then, changing mol Fe_2O_3 to mol Fe to grams Fe, we have

$0.0625 \; \cancel{mol \; Fe_2O_3} \times \dfrac{2 \; \cancel{mol \; Fe}}{1 \; \cancel{mol \; Fe_2O_3}} \times \dfrac{55.8 \; g \; Fe}{1 \; \cancel{mol \; Fe}} = 6.98 \; g \; Fe$

Example No. 8-10. In the previous example, how much Al will be left over?

Basing our calculations upon the limiting factor of 0.0625 mol Fe_2O_3 and changing this number of mol to mol of Al and then to grams of Al,

$0.0625 \; \cancel{mol \; Fe_2O_3} \times \dfrac{2 \; \cancel{mol \; Al}}{1 \; \cancel{mol \; Fe_2O_3}} \times \dfrac{27.0 \; g \; Al}{1 \; \cancel{mol \; Al}} = 3.38 \; g \; Al$

We have 4.00 g Al to begin with and we need 3.38 g, so that the excess is 4.00 g − 3.38 g Al or 0.62 g.

Example No. 8-11. 500 mL O_2 at 30°C and 740 torr will react with how many grams of Carbon to produce CO?

The reaction is $\qquad\qquad 2 \; C + O_2 \rightarrow 2 \; CO.$

Before we can use the weight-volume relationships as indicated by the equation, we must convert the volume of the O_2 to STP since these relationships hold true only at STP.

So, $500 \text{ mL} \times \frac{273 \text{ K}}{373 \text{ K}} \times \frac{740 \text{ torr}}{760 \text{ torr}} = 439 \text{ mL} = 0.439$ liters at STP.

Then, changing liters O_2 to moles O_2 to moles C to grams C, we have

$$0.439 \text{ L } O_2 \text{ STP} \times \frac{1 \text{ mol } O_2}{22.4 \text{ L } O_2 \text{ STP}} \times \frac{2 \text{ mol C}}{1 \text{ mol } O_2} \times \frac{12.0 \text{ g C}}{1 \text{ mol C}} = 0.470 \text{ g C}$$

Example No. 8-12. What volume of Cl_2 at 40°C and 755 torr will be produced by reacting 20 grams MnO_2 with excess HCl?

$$MnO_2 + 4 \text{ HCl} \rightarrow MnCl_2 + Cl_2 + 2 H_2O$$

Changing grams MnO_2 to moles MnO_2 to mol Cl_2 to liters Cl_2 at STP,

$$20 \text{ g } MnO_2 \times \frac{1 \text{ mol } MnO_2}{87 \text{ g } MnO_2} \times \frac{1 \text{ mol } Cl_2}{1 \text{ mol } MnO_2} \times \frac{22.4 \text{ L } Cl_2 \text{ STP}}{1 \text{ mol } Cl_2} = 5.2 \text{ L } Cl_2 \text{ STP}$$

However, the volume is to be given at 40°C and 755 torr so we must convert the volume at STP to a volume under the new conditions.

Thus, $5.2 \text{ liters} \times \frac{313 \text{ K}}{273 \text{ K}} \times \frac{760 \text{ torr}}{755 \text{ torr}} = 5.9$ liters at 40°C and 755 torr.

Problem Assignment

8-1.   How many grams of oxygen will be required to burn 0.157 g aluminum?

$4 \text{ Al} + 3 O_2 \rightarrow 2 Al_2O_3$ . . . . (0.140 g)

8-2.   What weight of CO_2 may be prepared from 500 g $CaCO_3$? What volume CO_2 at STP?

$CaCO_3 \rightarrow CaO + CO_2$ . . . . (220 g, 110 L)

8-3.   How many grams of hydrogen may be prepared from 5.00 g Al?

$2 \text{ Al} + 2 \text{ NaOH} + 6 H_2O \rightarrow 2 \text{ NaAl(OH)}_4 + 3 H_2$ . . . . (0.561 g)

8-4.   What volume will the hydrogen from problem 8-3 occupy at 20°C and 700 torr? . . . . . (7.20 L)

8-5.   How many grams of AgCl may be prepared from 1.00 liter of a solution containing 1.75 mg NaCl per mL?

$NaCl + Ag NO_3 \rightarrow AgCl + NaNO_3$ . . . . (4.30 g)

8-6.   What weight of oxygen will be required to burn 100 g coke (90% carbon)?

$C + O_2 \rightarrow CO_2$ . . . . (240 g)

8-7.   What volume of air (21.0% oxygen) at 30°C and 750 torr will be required to burn the coke in problem 8-6? . . . . (900 L)

8-8.   350 g SiO_2 are mixed with 60.0 g C and then heated to a high temperature. What volume of CO at 1000°C and one atmosphere pressure will be evolved? What weight SiC will be formed? . . . . (348 L, 66.8 g)

$$SiO_2 + 3\,C \rightarrow SiC + 2\,CO$$

8-9.   How many liters of oxygen at STP are required to burn 100 g sulfur? What volume SO_2 can be collected at 20°C and 740 torr?

8-10.   Assuming that the reaction $4\,NH_3 + 5\,O_2 \rightarrow 4\,NO + 6\,H_2O$ (all gases) goes to completion, from 150 mL NH_3,

        a. What volume NO will be produced?
        b. What volume steam will be produced?
        c. What volume oxygen will be needed?

8-11.   When a sample of BaO_2 was heated, 0.754 g of oxygen was evolved. What was the mass of the sample?

$$2\,BaO_2 \rightarrow 2\,BaO + O_2$$

8-12.   When a sample of mercuric oxide was heated, 468 mL O_2 were collected over water at 27°C and 754 torr. What was the mass of the sample?

$$2\,HgO \rightarrow 2\,Hg + O_2$$

8-13.   What volume CO at 1400°C and one atmosphere pressure will be required to react with 250 g Fe_2O_3?

$$Fe_2O_3 + 3\,CO \rightarrow 2\,Fe + 3\,CO_2$$

8-14.   20.0 g of iron are placed in 1.00 liter of a solution containing 40.0 mg $CuSO_4$ per mL. How much $FeSO_4$ is produced?

$$Fe + CuSO_4 \rightarrow FeSO_4 + Cu$$

8-15.   What 0.750 moles sugar, $C_6H_{12}O_6$, are burned in oxygen, how many moles CO_2 are formed? How many moles H_2O are formed? How many mL of water are formed?

$$C_6H_{12}O_6 + 6\,O_2 \rightarrow 6\,CO_2 + 6\,H_2O$$

8-16.   What weight of $BaSO_4$ will be precipitated from 250 mL of a solution containing 20.0 mg $Ba(NO_3)_2$ per mL if 10.0 g Na_2SO_4 are added?

$$Ba(NO_3)_2 + Na_2SO_4 \rightarrow BaSO_4 + 2\,NaNO_3$$

8-17.   If 11.2 liters of hydrogen at STP are reacted with 45.0 g CuO, how many grams of water will be produced?

$$H_2 + CuO \rightarrow Cu + H_2O$$

8-18.   What mass CO_2 will be produced upon the complete combustion of 10.0 lb. octane?

$$2\,C_8H_{18} + 25\,O_2 \rightarrow 16\,CO_2 + 18\,H_2O$$

8-19.   A 15.0 g sample of $KClO_3$ is heated strongly. How much O_2 is evolved? The solid product is dissolved in 100 mL of water containing 20.0 g $AgNO_3$. What and how much are the products?

8-20.   How much iron can be produced from 100 tons of ore which is 70% Fe_3O_4 if the process if 68% efficient? How much 95% pure iron can be obtained?

$$Fe_3O_4 + 4\ CO \rightarrow 3\ Fe + 4\ CO_2$$

8-21.   An impure piece of zinc, (weighing 0.500 g) when treated with hydrochloric acid yielded 100 mL H_2 collected over water at 15°C and 755 torr. What is the percentage purity of the zinc?

8-22.   During the electrolysis of molten NaCl, 27.5 liters Cl_2 at 600°C and one atmosphere pressure are produced. What mass Na was formed? What mass of NaCl was decomposed?

8-23.   A sample containing 60.0 g carbon is burned in the presence of oxygen. The excess oxygen is then used to burn 40.0 g sulfur. The products of both of these reactions are then dissolved in water. How much H_2CO_3 and H_2SO_3 are formed? If 250 g of oxygen were available initially, how much oxygen would be left over?

8-24.   From the reaction $4\ Fe(CrO_2)_2 + 8\ K_2CO_3 + 7\ O_2 \rightarrow 2\ Fe_2O_3 + 8\ K_2CrO_4 + 8\ CO_2$ calculate:

a. the number of moles K_2CrO_4 produced from 5.00 moles $Fe(CrO_2)_2$
b. the number of moles $Fe(CrO_2)_2$ required to produce 3.00 moles Fe_2O_3
c. the number of moles $Fe(CrO_2)_2$ required to produce 3.00 g Fe_2O_3
d. the number of moles $Fe(CrO_2)_2$ required to produce 3.00 moles CO_2
e. the number of moles $Fe(CrO_2)_2$ required to produce 3.00 g CO_2
f. the number of moles $Fe(CrO_2)_2$ required to produce 3.00 L CO_2 at STP
g. the number of moles $Fe(CrO_2)_2$ required to react with 2.00 moles O_2
h. the number of moles $Fe(CrO_2)_2$ required to react with 4.00 moles K_2CO_3
i. the number of grams $Fe(CrO_2)_2$ required to react with 1.00 g O_2
j. the number of grams $Fe(CrO_2)_2$ required to react with 5.00 L O_2 STP

8-25.   An automobile engine burns 0.060 gallons of gasoline for each mile travelled. How much gasoline will be required for a 200 mile trip? How much air (21% oxygen) at 70° F and one atmosphere pressure will be required to burn this amount of gasoline, assuming it to be isooctane, C_8H_{18}, density 0.69 g/mL. (1 gal = 3.79 L).

$$2\ C_8H_{18} + 25\ O_2 \rightarrow 16\ CO_2 + 18\ H_2O$$

8-26.   Calculate the mass of $NaNO_3$ produced from 68.0 g NH_3 according to the following reactions:

$$4\ NH_3 + 5\ O_2 \rightarrow 4\ NO + 6\ H_2O$$
$$2\ NO + O_2 \rightarrow 2\ NO_2$$
$$2\ NO_2 + H_2O \rightarrow HNO_3 + HNO_2$$
$$HNO_3 + NaOH \rightarrow NaNO_3 + H_2O$$

8-27.   For the reaction $C_{12}H_{22}O_{11} + 12\ O_2 \rightarrow 12\ CO_2 + 11\ H_2O$ calculate

   a. the number of moles CO_2 produced from 0.500 moles $C_{12}H_{22}O_{11}$
   b. the number of moles O_2 required to produce 3.40 moles H_2O.
   c. the number of moles O_2 required to react with 1.20 moles $C_{12}H_{22}O_{11}$.
   d. the number of moles O_2 required to react with 1.20 g $C_{12}H_{22}O_{11}$.
   e. the number of moles $C_{12}H_{22}O_{11}$ required to react with 1.20 L O_2 at STP.
   f. the number of moles $C_{12}H_{22}O_{11}$ required to react with 1.20 g O_2.
   g. the number of moles $C_{12}H_{22}O_{11}$ required to produce 5.00 L H_2O at 100°C and 745 torr.
   h. the number of grams $C_{12}H_{22}O_{11}$ required to produce 5.00 g CO_2.
   i. the number of grams $C_{12}H_{22}O_{11}$ required to produce 5.00 L CO_2 at STP.

8-28.   A 5.00 g mixture of $MgCO_3$ and $CaCO_3$ is heated, yielding 2.36 g CO_2. What is the percent of $MgCO_3$ in the mixture?

8-29.   When 3.08×10^{-4} mL hydrogen at 20°C and 1.00×10^{-4} torr are mixed with an excess of oxygen, how many mL of (liquid) water will be formed? How many moles of water will be formed? How many molecules?

8-30.   A 55.00 mg mixture of zinc and calcium reacts with acid to liberate 23.37 mL of hydrogen collected over water at 20° C and 760 torr. What is the percent composition of the mixture?

Chapter 9

ELECTRONIC STRUCTURE

Before we begin our discussion of the actual electronic structures of the elements, we should review some general facts about atomic structure.

I. ATOMIC STRUCTURE

All atoms are composed of a nucleus with electrons outside that nucleus.

An atom consists of three fundamental particles—the electron, the proton, and the neutron. We can summarize their main physical properties in the following table:

Name of particle	Symbol	Charge	Approximate mass in atomic mass units	Location
Electron	e^-	-1	0.00055	outside nucleus
Proton	p	$+1$	1	inside nucleus
Neutron	n	0	1	inside nucleus

The nucleus of an atom contains the protons and the neutrons, while the electrons are located outside the nucleus. To determine the number of each of these particles present in any given atom, we observe the following rules:

1. The number of protons in the nucleus is always equal to the atomic number of the element in question.
2. The number of electrons is always equal to the number of protons since the positive and negative charges must always neutralize each other.
3. The number of neutrons is equal to the difference between the atomic number and the mass number. (This is based on the assumption that all of the mass of the atom is located in the nucleus.)

A few examples will clear up these points.

Example No. 9-1.   Let's take Hydrogen, symbol H, atomic number 1, mass number 1. The atomic number is 1, so there must be 1 proton and 1 electron. Thus we can draw the structure

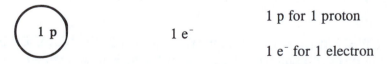

1 p for 1 proton

1 e⁻ for 1 electron

The circle represents the nucleus with the proton inside and the electron outside. Next, to determine the number of neutrons, we take the difference between the atomic number (1) and the mass number (1) which is zero (0). This represents the number of neutrons, so we write

0 n for zero neutrons

This represents the atomic structure of Hydrogen.

Example No. 9-2. Now let's take Helium, symbol He, atomic number 2, mass number 4. Since the atomic number is 2, there must be 2 protons in the nucleus and 2 electrons outside the nucleus, or

2 e⁻

The difference between atomic number (2) and mass number (4) is 2 so there must be 2 neutrons in the nucleus. Therefore, Helium has the structure

2 e⁻

Example No. 9-3. Next, let us take a more complicated example. Sodium, symbol Na, has an atomic number of 11 and mass number of 23. The atomic number tells us that there are 11 protons in the nucleus and 11 electrons outside the nucleus, or

11 e⁻

The difference between the atomic number (11) and the mass number (23) is 12, so there must be 12 neutrons in the nucleus. Thus, the structure for Sodium is

11 e⁻

Example No. 9-4. In the case of Uranium, symbol U, atomic number 92 and mass number 238, there must be 92 protons and 92 electrons and 238 minus 92 or 146 neutrons, so we have

92 e⁻

II. ISOTOPES

In looking at the table of atomic masses of the elements, you have probably noticed what while most of these masses are nearly whole numbers, some of them are not. This is due to the presence of isotopes. Let us define this term and then see how it fits into the above discussion.

ISOTOPES ARE ATOMS OF AN ELEMENT WITH THE SAME ATOMIC NUMBER BUT DIFFERENT MASS NUMBERS. From this definition we can see that since the atomic numbers are the same, the elements must have the same number of protons and electrons. Since the mass numbers are different, the number of neutrons must be different in each case. The best way to understand this is to try a few problems and then go back to the definition and reevaluate it.

Example No. 9-5 Take Chlorine, symbol Cl, atomic number 17 and mass numbers 35 and 37 (from list of isotopes). Each isotope must have 17 protons and 17 electrons since the atomic number is 17, or

17 e⁻   17 e⁻

Since one has a mass number of 35 it must have 35-17 or 18 neutrons; the other with a mass number of 37 must have 37-17 or 20 neutrons. Thus for the two isotopes we have

The *atomic mass* is the average mass of all of the isotopes. If the two isotopes of chlorine, mass numbers 35 and 37, were present in equal amounts, the atomic (average) mass would be 36. However, since the atomic mass is listed as 35.453, the isotope of mass number 35 must be the predominant one, because the atomic mass is closer to 35 than to 37.

Example No. 9-6.   Let's take Hydrogen, symbol H, with atomic number 1 and mass numbers 1, 2, and 3. We know that each must have 1 proton and 1 electron since the atomic number of each is 1 so we have

For the first isotope we find that there are no neutrons (by subtracting the atomic number 1 from the mass number 1); for the second isotope, there must be 1 neutron (the atomic number 1 from the mass number 2); there must be 2 neutrons in the third isotope (atomic number 1 subtracted from mass number 3) so that the structure for the three isotopes will be

From the chart of atomic masses we see that Hydrogen is listed at 1.008. This tells us that the Hydrogen of mass number 1 is the most common, with that of mass number 2 occurring in smaller amounts and that of mass number 3 in very minute amounts.

III.   ELECTRON ENERGY LEVELS

Electrons may exist in several different energy levels around the nucleus. These energy levels are numbered 1, 2, 3, 4—and are indicated by the letter n. The lowest energy level, $n = 1$, is that energy level closest to the nucleus.

The maximum number of electrons possible in each energy level may be calculated by using the formula $2n^2$ where n is the number of that energy level. For $n = 1$, $2n^2 = 2(1)^2 = 2$ so that the first energy level can hold a maximum of 2 electrons. It can hold 1 electron or 2 electrons but never more than 2. For $n = 2$, $2n^2 = 2(2)^2 = 8$. That is, the 2nd energy level can hold any number of electrons up to and including 8 but never more than 8. The third energy level can hold a maximum of 18 electrons ($2n^2 = 2(3)^2 = 18$). Likewise, the 4th energy level can hold a maximum of 32 electrons ($2n^2 = 2(4)^2 = 32$).

Thus we have the following energy levels showing the maximum number of electrons each can contain.

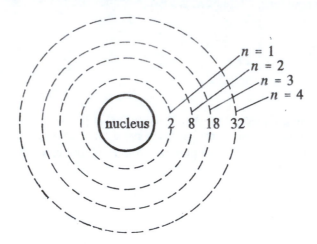

THE 1st ENERGY LEVEL MUST BE FILLED BEFORE WE CAN BEGIN TO PLACE ELECTRONS IN THE 2nd ENERGY LEVEL; THE 2nd ENERGY LEVEL MUST BE FILLED BEFORE WE CAN START PLACING ELECTRONS IN THE 3rd ENERGY LEVEL.

A few examples will clear up these points much better than any explanation, so let us draw the electronic structure of the elements starting with Hydrogen, atomic number 1, letting a circle with the symbol inside represent the nucleus of the element.

Example No. 9-7.   Hydrogen, atomic number 1, has only 1 electron. It must go into the first energy level, so we have

We will use a curved line to represent an energy level to save space and time in writing. But we should remember that it represents a complete energy level.

Example No. 9-8.   Helium, symbol He, has an atomic number of 2, so it has 2 electrons. Both of these can go into the 1st energy level since the 1st energy level can hold 2 electrons, so we have

Example No. 9-9.   Lithium, symbol Li, atomic number 3, has 3 electrons. The 1st energy level can hold only 2 electrons so the third must go into the 2nd energy level, or

Example No. 9-10.   Beryllium, symbol Be, has an atomic number of 4, and so contains 4 electrons. The 1st energy level can hold only 2, so the remaining 2 must go into the 2nd energy level, or

Example No. 9-11. Fluorine, symbol F, atomic number 9, has 9 electrons outside the nucleus. Two of these can go into the 1st energy level and the remaining 7 into the 2nd energy level since the 2nd energy level can hold up to 8 electrons, or

Example No. 9-12. Sodium, symbol Na, atomic number 11, has 11 electrons outside the nucleus. The 1st energy level can hold 2 of them, the 2nd energy level can hold only 8 so that the eleventh electron must go into the 3rd energy level, or

$$\text{Na} \quad 2 \Big) \quad 8 \Big) \quad 1 \Big)$$

Example No. 9-13. Magnesium, symbol Mg, atomic number 12, will have 2 electrons in the 1st energy level, 8 in the 2nd energy level and 2 in the 3rd energy level, or

$$\text{Mg} \quad 2 \Big) \quad 8 \Big) \quad 2 \Big)$$

Example No. 9-14. Argon, symbol Ar, atomic number 18, will have 2 electrons in the 1st energy level, 8 electrons in the 2nd energy level, and 8 electrons in the 3rd energy level, or

$$\text{Ar} \quad 2 \Big) \quad 8 \Big) \quad 8 \Big)$$

We know from spectroscopic data that all elements tend to reach the rare gas structure which is the most stable one. Therefore, we know that an outer energy level of 8 electrons will be the most stable configuration and that the atoms will try to keep this structure as long as possible.

Example No. 9-15. Thus, for Potassium, symbol K, atomic number 19, we would have 2 electrons in the 1st energy level, 8 in the 2nd energy level, and 9 electrons to place elsewhere. We know that the 3rd energy level can hold up to 18 electrons but 8 is a very stable structure, so that we actually find in Potassium 8 electrons in the 3rd energy level and 1 electron in the 4th energy level, or

$$\text{K} \quad 2 \Big) \quad 8 \Big) \quad 8 \Big) \quad 1 \Big)$$

Example No. 9-16. For Calcium, symbol Ca, atomic number 20, we have 2 electrons in the 1st energy level, 8 in the 2nd energy level, 8 in the 3rd energy level, and 2 in the 4th energy level, or

$$\text{Ca} \quad 2 \Big) \quad 8 \Big) \quad 8 \Big) \quad 2 \Big)$$

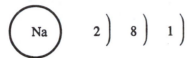

IV. TRANSITION ELEMENTS

The above general rules hold true until we come to the transition elements where the electrons do not always complete one energy level before starting another. Let us define the term transition element and give a few examples. Then a review of the definition should clear up any remaining questions.

TRANSITION ELEMENTS ARE THOSE WHOSE INNER ELECTRON ENERGY LEVELS ARE BEING FILLED.

Example No. 9-17. Scandium, symbol Sc, atomic number 21 is the first transition element. We know that we have 2 electrons in the 1st energy level, 8 in the 2nd energy level, and 11 more to place elsewhere. Another general rule that we can now use is that most of the transition elements have 2 electrons in their outermost energy level. Please note the word "most"; the exceptions may be explained in a more advanced text. Therefore, since we should have 2 electrons in the outermost energy level, we can write.

$$\left(\text{Sc}\right) \quad 2\Big) \quad 8\Big) \quad ?\Big) \quad 2\Big)$$

This gives us a total of 12 electrons so far, leaving 9 for the 3rd energy level, or

$$\left(\text{Sc}\right) \quad 2\Big) \quad 8\Big) \quad 9\Big) \quad 2\Big)$$

The 3rd energy level can hold up to 18 electrons, so we see how the definition works; an inner electron energy level is now being filled.

Example No. 9-18. Titanium, symbol Ti, atomic number 22, will have 2 electrons in the 1st energy level shell, 8 in the 2nd energy level, and 2 in the 4th energy level, or

$$\left(\text{Ti}\right) \quad 2\Big) \quad 8\Big) \quad ?\Big) \quad 2\Big)$$

This leaves a balance of 10 electrons which can go into the 3rd energy level giving the structure

$$\left(\text{Ti}\right) \quad 2\Big) \quad 8\Big) \quad 10\Big) \quad 2\Big)$$

Example No. 9-19. Cobalt, symbol Co, atomic number 27 will have 2 electrons in the 1st energy level, 8 in the 2nd energy level, and 2 in the 4th energy level, or

$$\left(\text{Co}\right) \quad 2\Big) \quad 8\Big) \quad ?\Big) \quad 2\Big)$$

This leaves a balance of 15 electrons for the 3rd energy level so we have

$$\text{Co} \quad 2 \big) \; 8 \big) \; 15 \big) \; 2 \big)$$

After the 3rd energy level has a total of 18 electrons, the 4th energy level begins to fill up until it has a total of 8 electrons. Then, after there are 2 electrons in the 5th energy level, the 4th energy level begins to fill again, thus starting a new transition series.

Example No. 9-20. Zinc, symbol Zn, atomic number 30, has 2 electrons in the 1st energy level, 8 in the 2nd energy level, 18 in the 3rd energy level, and 2 in the 4th energy level, or

$$\text{Zn} \quad 2 \big) \; 8 \big) \; 18 \big) \; 2 \big)$$

Example No. 9-21. For Krypton, symbol Kr, atomic number 36, there are 2 electrons in the 1st energy level, 8 in the 2nd energy level, 18 in the 3rd energy level, and 8 in the 4th energy level, or

$$\text{Kr} \quad 2 \big) \; 8 \big) \; 18 \big) \; 8 \big)$$

Now let us repeat the definition of transition elements—ELEMENTS WHOSE INNER ENERGY LEVELS OF ELECTRONS ARE BEING FILLED. Thus:

$$\text{Sc} \quad 2 \big) \; 8 \big) \; 9 \big) \; 2 \big) \qquad \text{is a transition element}$$

$$\text{Ti} \quad 2 \big) \; 8 \big) \; 10 \big) \; 2 \big) \qquad \text{is a transition element}$$

$$\text{Ni} \quad 2 \big) \; 8 \big) \; 16 \big) \; 2 \big) \qquad \text{is a transition element}$$

$$\text{Kr} \quad 2 \big) \; 8 \big) \; 18 \big) \; 8 \big) \qquad \text{is } not \text{ a transition element}$$

V. QUANTUM NUMBERS

On the basis of experiments and by means of quantum mechanical calculations, it can be shown that an electron energy level (also called a major energy level) consists of one or more sublevels. There are four basic types of energy levels and sublevels which in turn are designated by four quantum numbers.

1. *The principal quantum number, n,* describes the energy level an electron occupies. n may have any integral value: 1,2,3,4. . . . That is, $n=1$ describes the lowest energy level, the one closest to the nucleus.

2. *The azimuthal quantum number,* ℓ, designates the geometric shape of the region in space that an electron occupies. The value of ℓ ranges from 0 to and including $(n-1)$. That is:

for $n=1$, $\ell=0$

$n=2$, $\ell=0$, 1

$n=3$, $\ell=0$, 1, 2.

The azimuthal quantum number designates an energy sublevel, or a specific kind of atomic orbital, that an electron can occupy.

Note that the number of energy sublevels described by the azimuthal quantum number equals the number of the main energy level (n). These energy sublevels, ℓ, are designated by letters s, p, d, f, g . . . as follows:

$\ell=0$	$\ell=1$	$\ell=2$	$\ell=3$
s	p	d	f

Thus "s" represents the lowest energy sublevel that an electron may occupy; "p" the next sublevel, etc.

Combining this information, we have:

For $n=1$, $\ell=0$. That is, the first main energy level has only one energy sublevel which may be designated as $1s$, the "s" sublevel in the first main energy level.

For $n=2$, $\ell=0,1$. That is, the second main energy level has two sublevels, the $2s$ and the $2p$.

For $n=3$, $\ell=0,1,2$. The third main energy level has three sublevels, $3s$, $3p$, and $3d$. Likewise the fourth main energy level has four sublevels, $4s$, $4p$, $4d$, $4f$.

The relationships between the main energy levels and the sublevels for different values of n may be shown as:

$n=1$ . . . . . . $1s$

$n=2$ . . . . . . $2p$
. . . . $2s$

. . . . $3d$
$n=3$ . . . . . . $3p$
. . . . $3s$

. . . . $4f$
. . . . . . $4d$
$n=4$ . . . . . . $4p$
. . . . $4s$

3. *The magnetic quantum number,* m_ℓ, designates the spatial arrangement of an atomic orbital.

m_ℓ has integral values of $-\ell$ to 0 to $+\ell$. Each value of m_ℓ corresponds to an atomic orbital, or a region in space where there is a high probability of finding an electron. Thus, when $n=1$, $\ell=0$ and $m_\ell=0$, there is only one value of m_ℓ so that there is only one $1s$ atomic orbital possible. When $n=2$, ℓ can have values of 0 and 1. Let us consider each value of ℓ separately.

When $n=2$, and $\ell=0$, $m_\ell=0$ so that there is only one s atomic orbital, the $2s$; when $n=2$, and $\ell=1$, $m_\ell=-1$, 0, $+1$. There are three values of m_ℓ so that there are three types of atomic orbitals possible. These are designated as $2p_x$, $2p_y$, $2p_z$ since a value of $\ell=1$ corresponds to a "p" sublevel. Each of the "p" atomic orbitals has the same energy and differs primarily in the spatial orientation of the electron clouds.

4. *The spin quantum number, m_s,* refers to the spin of an electron and the orientation of the magnetic field produced by that spinning electron. m_s can have values $\pm 1/2$.

The quantum numbers n, ℓ, m_ℓ describe a particular atomic orbital. Each atomic orbital can accommodate no more than two electrons, one with $m_s = +1/2$ and one with $m_s = -1/2$.

QUANTUM NUMBERS, ENERGY LEVELS AND SUBLEVELS, AND NUMBERS OF ELECTRONS

n	ℓ	m_ℓ	m_s	simplified designation	number of electrons in sublevels	total number of electrons in energy level
1	0	0	$\pm 1/2$	$1s$	2	2
2	0	0	$\pm 1/2$	$2s$	2	
	1	$-1, 0, +1$	$\pm 1/2$	$2p_x$, $2p_y$, $2p_z$	6	8
3	0	0	$\pm 1/2$	$3s$	2	
	1	$-1, 0, +1$	$\pm 1/2$	$3p_x$, $3p_y$, $3p_z$	6	18
	2	$-2, -1, 0, +1, +2$	$\pm 1/2$	five $3d$ atomic orbitals	10	

VI. ARRANGEMENT OF ELECTRONS IN ATOMIC ORBITALS

In this chapter we will use only the first two quantum numbers, n and ℓ, in showing the electronic structure of several different atoms. The following chart shows the maximum number of atomic orbitals available. Note that a given atomic orbital, designated below by a box, may contain no more than 2 electrons (designated by arrows). That atomic orbital may contain 1 electron, or none, but never more than 2.

ℓ sublevel		maximum number of electrons
f	⊡ ⊡ ⊡ ⊡ ⊡ ⊡ ⊡	14
d	⊡ ⊡ ⊡ ⊡ ⊡	10
p	⊡ ⊡ ⊡	6
s	⊡	2

If the "s" atomic orbital is filled, it can hold 2 electrons; if "s and p" atomic orbitals are filled, they can hold a maximum of 8 electrons; if "s, p, and d" atomic orbitals are filled, the maximum number of electrons is 18; and, if "s, p, d, f" atomic orbitals are filled, the maximum number of electrons is 32.

The number of electrons in a given atomic orbital of a given energy level may be indicated as

$$3s^2$$

where the first number indicates the main energy level, the letter the type of atomic orbital or energy sublevel, and the superscript the number of electrons present in that atomic orbital. Thus the above representation indicates 2 electrons in the "s" atomic orbital of the 3rd main energy level.

The orbitals are filled in a very definite sequence as indicated in the following table:

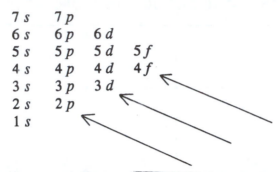

The order of filling the orbitals is found by reading diagonally upward in the direction indicated by the arrows. Thus the orbitals are filled in the order: $1s$, $2s$, $2p$, $3s$, $3p$, $4s$, $3d$, $4p$, etc. We might think that it would be easier simply to sit down and memorize this sequence but after a little practice we will begin to remember in what order the orbitals are filled.

Example No. 9-22. For Hydrogen, atomic number 1, we have only one electron. This electron must fill the lowest energy level, the $1s$ atomic orbital, first, so that for hydrogen we write $1s^1$, which means one electron in the s atomic orbital of the first energy level.

Example No. 9-23. For Helium, atomic number 2, we have two electrons. These can both go into the $1s$ atomic orbital since this atomic orbital can hold a total of two electrons. Thus for Helium we write ⊙He $1s^2$, which indicates two electrons in the s orbital of the first energy level.

Example No. 9-24. For Lithium, atomic number 3, we have a total of three electrons. The $1s$ atomic orbital can hold only 2 electrons or $1s^2$. The remaining electron must go into the next available atomic orbital, the $2s$ atomic orbital, so that the electron arrangement for Lithium is $1s^2$ $2s^1$, where the $2s^1$ indicates 1 electron in the s atomic orbital of the second energy level. We should note that this agrees with our previous discussion of the structure of Lithium as ⊙Li $2)$ $1)$.

Example No. 9-25. For Beryllium, atomic number 4, we have a total of 4 electrons to place in the proper atomic orbitals. The first atomic orbital, the $1s$, can hold only 2 electrons, or $1s^2$. The next atomic orbital, the $2s$ atomic orbital can also hold 2 electrons or $2s^2$ so that the structure of Beryllium is

Be . . . . . $1s^2\ 2s^2$

Example No. 9-26. For Boron, atomic number 5, we have 5 electrons. The first atomic orbital, the $1s$, can hold 2 electrons or $1s^2$. The second atomic orbital, the $2s$, can also hold 2 electrons or $2s^2$. The fifth electron must then go into the next available atomic orbital, the $2p$, or $2p^1$, so that for Boron, we have

$1s^2\ 2s^2\ 2p^1$

Example No. 9-27. Next let us go on to Fluorine, atomic number 9 with 9 electrons to be placed in the proper atomic orbitals. Again we will place 2 electrons in the $1s$ atomic orbital and also 2 electrons in the $2s$ atomic orbital since this is all the electrons these atomic orbitals can hold. The remaining 5 electrons must go into the $2p$ atomic orbitals since there are 3 of these $2p$ atomic orbitals and each atomic orbital can hold 2 electrons. Thus the structure for Fluorine is $1s^2\ 2s^2\ 2p^5$.

Example No. 9-28. For Magnesium, atomic number 12, we have 2 electrons in the $1s$ atomic orbital, and 2 electrons in the $2s$ atomic orbital. The next atomic orbitals, the $2p$, can hold 6 electrons, giving us a total of 10 electrons so far. The remaining 2 electrons must go into the next atomic orbital, the $3s$ atomic orbital, so that the structure for magnesium is

$$1s^2\ 2s^2\ 2p^6\ 3s^2$$

Example No. 9-29. For Potassium, atomic number 19, we have 19 electrons to place in the atomic orbitals. Again we can place 2 electrons in the $1s$ atomic orbital, and 2 electrons in the $2s$ atomic orbital. The $2p$ atomic orbitals can hold 6 electrons making a total of 10 electrons so far. The next atomic orbital to be filled is the $3s$ which can hold only 2 electrons. Then come the $3p$ atomic orbitals which can hold a total of 6 electrons, making a total of 18 electrons. The one remaining electron must go into the next atomic orbital in order, the $4s$ atomic orbital, so that the structure of Potassium is

$$1s^2\ 2s^2\ 2p^6\ 3s^2\ 3p^6\ 4s^1$$

Example No. 9-30. If we consider a transition element such as cobalt, atomic number 27, we can fill the atomic orbitals in the same manner. The $1s$ atomic orbital can hold 2 electrons or $1s^2$. The $2s$ atomic orbital can hold 2 electrons, $2s^2$, and the $2p$ atomic orbitals can hold 6 electrons or $2p^6$. The $3s$ atomic orbital can hold 2 electrons or $3s^2$ and the $3p$ atomic orbitals can hold 6 electrons or $3p^6$. This gives us a total of 18 electrons with 9 more to be placed in their proper atomic orbitals. The next atomic orbital in order is the $4s$ atomic orbital, which can hold only 2 electrons or $4s^2$. The remaining 7 electrons go into the next atomic orbitals, the $3d$ atomic orbitals, which can hold up to 10 electrons or $3d^7$. Thus for Cobalt we have

$$1s^2\ 2s^2\ 2p^6\ 3s^2\ 3p^6\ 4s^2\ 3d^7$$

We should note that this agrees with our previous discussion of the structure of Cobalt as 2)8)15)2).

THE PERIODIC TABLE
List of Outer Electrons Only

	IA (1)	IIA (2)	IIIB (3)	IVB (4)	VB (5)	VIB (6)	VIIB (7)	VIIIB (8)	VIIIB (9)	VIIIB (10)	IB (11)	IIB (12)	IIIA (13)	IVA (14)	VA (15)	VIA (16)	VIIA (17)	VIIIA (18)
	I	II	III	IV	V	VI	VII	VIII	VIII	VIII			III	IV	V	VI	VII	VIII
1	H $1s^1$																	He $1s^2$
2	Li $2s^1$	Be $2s^2$											B $2s^2\,2p^1$	C $2s^2\,2p^2$	N $2s^2\,2p^3$	O $2s^2\,2p^4$	F $2s^2\,2p^5$	Ne $2s^2\,2p^6$
3	Na $3s^1$	Mg $3s^2$											Al $3s^2\,3p^1$	Si $3s^2\,3p^2$	P $3s^2\,3p^3$	S $3s^2\,3p^4$	Cl $3s^2\,3p^5$	Ar $3s^2\,3p^6$
4	K $4s^1$	Ca $4s^2$	Sc $4s^2\,3d^1$	Ti $4s^2\,3d^2$	V $4s^2\,3d^3$	Cr $4s^1\,3d^5$	Mn $4s^2\,3d^5$	Fe $4s^2\,3d^6$	Co $4s^2\,3d^7$	Ni $4s^2\,3d^8$	Cu $4s^1\,3d^{10}$	Zn $4s^2\,3d^{10}$	Ga $4s^2\,4p^1\,3d^{10}$	Ge $4s^2\,4p^2\,3d^{10}$	As $4s^2\,4p^3\,3d^{10}$	Se $4s^2\,4p^4\,3d^{10}$	Br $4s^2\,4p^5\,3d^{10}$	Kr $4s^2\,4p^6\,3d^{10}$
5	Rb $5s^1$	Sr $5s^2$	Y $5s^2\,4d^1$	Zr $5s^2\,4d^2$	Nb $5s^1\,4d^4$	Mo $5s^1\,4d^5$	Tc $5s^2\,4d^5$	Ru $5s^1\,4d^7$	Rh $5s^1\,4d^8$	Pd $5s^0\,4d^{10}$	Ag $5s^1\,4d^{10}$	Cd $5s^2\,4d^{10}$	In $5s^2\,5p^1\,4d^{10}$	Sn $5s^2\,5p^2\,4d^{10}$	Sb $5s^2\,5p^3\,4d^{10}$	Te $5s^2\,5p^4\,4d^{10}$	I $5s^2\,5p^5\,4d^{10}$	Xe $5s^2\,5p^6\,4d^{10}$
6	Cs $6s^1$	Ba $6s^2$	La* $6s^2\,5d^1$	Hf $6s^2\,5d^2\,4f^{14}$	Ta $6s^2\,5d^3\,4f^{14}$	W $6s^2\,5d^4\,4f^{14}$	Re $6s^2\,5d^5\,4f^{14}$	Os $6s^2\,5d^6\,4f^{14}$	Ir $6s^2\,5d^7\,4f^{14}$	Pt $6s^1\,5d^9\,4f^{14}$	Au $6s^1\,5d^{10}\,4f^{14}$	Hg $6s^2\,5d^{10}\,4f^{14}$	Tl $6s^2\,6p^1\,5d^{10}\,4f^{14}$	Pb $6s^2\,6p^2\,5d^{10}\,4f^{14}$	Bi $6s^2\,6p^3\,5d^{10}\,4f^{14}$	Po $6s^2\,6p^4\,5d^{10}\,4f^{14}$	At $6s^2\,6p^5\,5d^{10}\,4f^{14}$	Rn $6s^2\,6p^6\,5d^{10}\,4f^{14}$
7	Fr $7s^1$	Ra $7s^2$	Ac** $7s^2\,6d^1$	**														

* Lanthanides

VB	VIB	VIIB	(8)	(9)	(10)	IB	IIB	IIIA	IVA	VA	VIA	VIIA	VIIIA
Ce $6s^2\,5d^0\,4f^2$	Pr $6s^2\,5d^0\,4f^3$	Nd $6s^2\,5d^0\,4f^4$	Pm $6s^2\,5d^0\,4f^5$	Sm $6s^2\,5d^0\,4f^6$	Eu $6s^2\,5d^0\,4f^7$	Gd $6s^2\,5d^1\,4f^7$	Tb $6s^2\,5d^0\,4f^9$	Dy $6s^2\,5d^0\,4f^{10}$	Ho $6s^2\,5d^0\,4f^{11}$	Er $6s^2\,5d^0\,4f^{12}$	Tm $6s^2\,5d^0\,4f^{13}$	Yb $6s^2\,5d^0\,4f^{14}$	Lu $6s^2\,5d^1\,4f^{14}$

** Actinides

VB	VIB	VIIB	(8)	(9)	(10)	IB	IIB	IIIA	IVA	VA	VIA	VIIA	VIIIA
Th $7s^2\,6d^2\,5f^0$	Pa $7s^2\,6d^1\,5f^2$	U $7s^2\,6d^1\,5f^3$	Np $7s^2\,6d^1\,5f^4$	Pu $7s^2\,6d^1\,5f^5$	Am $7s^2\,6d^0\,5f^7$	Cm $7s^2\,6d^1\,5f^7$	Bk $7s^2\,6d^0\,5f^8$	Cf $7s^2\,6d^1\,5f^9$	Es $7s^2\,6d^1\,5f^{10}$	Fm $7s^2\,6d^1\,5f^{11}$	Md $7s^2\,6d^1\,5f^{12}$	No $7s^2\,6d^0\,5f^{14}$	Lr $7s^2\,6d^1\,5f^{14}$

VI.   THE NOBLE GASES

Now we can see why the noble gases are so unreactive. Let us make a chart of their electron energy levels and see how they compare with one another.

Symbol	Atomic Number	$n=1$	$n=2$	$n=3$	$n=4$	$n=5$	$n=6$
He	2	2					
Ne	10	2	8				
Ar	18	2	8	8			
Kr	36	2	8	18	8		
Xe	54	2	8	18	18	8	
Rn	86	2	8	18	32	18	8

We notice that all of these elements have a complete outer energy level of 8 electrons except Helium which requires only 2 to fill its outer energy level. This then is the most stable structure possible and all other elements will tend to approach this type of structure in order to become more stable. Also we will note below that the subenergy levels are the most stable when they are completely full as above, or when they are *half* full.

VII.   IONIZATION ENERGY

The ionization energy of an element is the amount of energy that must be added to an atom to remove an electron from its highest energy level. Let us study the chart below and see if we can correlate atomic structure with the ionization energies.

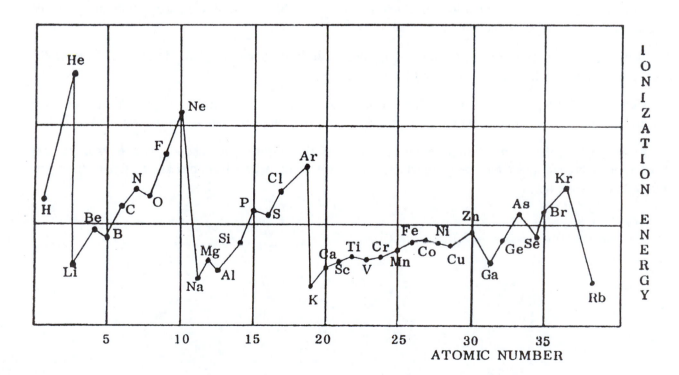

Hydrogen has only 1 electron outside the nucleus and it will take a certain amount of energy to remove this electron. Helium has 2 electrons in its first energy level which makes this energy level complete and thus very stable so that it should take more energy to remove one electron from this stable structure than from Hydrogen and we can see that it does because the ionization energy goes up from H to He on the preceding chart.

Lithium has 2 electrons in its first energy level and one in the second. It will tend to lose the electron from the second energy level so that it can have the inert gas structure like He and thus be more stable. Therefore, we would expect this electron to be lost easily since it is in another energy level farther from the nucleus. On the chart we see that Li actually does have a much lower ionization energy than He.

Beryllium has 2 electrons in its second energy level thus completely filling the $2s$ atomic orbital. This should make it more stable than Li and it should take more energy to remove an electron. Also the greater positive charge in the nucleus should hold these electrons more closely and we would expect the ionization energy to go up from Li to Be as it does.

Boron has 3 electrons in the 3rd energy level or $2s^2\,2p^1$. The p electron is in a slightly different energy level than the s electrons and so is not held as tightly and should be removed more easily. So we see why we get a drop in the ionization energy from Be to B.

Carbon has 4 electrons in the 3rd energy level or $2s^2\,2p^2$. The p electrons are held more tightly than in B since the positive nuclear charge is greater and the atomic orbital energy remains the same so the ionization energy goes up.

In the case of N, $2s^2\,2p^3$, the ionization energy goes up again for the same reason. However we note that the $2p$ atomic orbitals are half full, which is a stable structure.

Oxygen has the structure $2s^2\,2p^4$. It will tend to lose the fourth electron in the p sublevel so as to approach the stability of a sublevel half full. Therefore, we would expect a drop in the ionization energy when going from N to O because of this tendency to lose the extra electron in the p sublevel.

Fluorine, $2s^2\,2p^5$ is similar to Carbon in that the positive nuclear charge increases while the atomic orbital energy remains the same. It cannot lose an electron to have a stable structure so its ionization energy should be greater than that of Oxygen and we see that it is.

Neon has a very stable structure since its second energy level is completely full. Therefore, its ionization energy should be greater than that of Fluorine and it is.

Sodium, $2s^2\,2p^6\,3s^1$, is similar to Lithium. It will easily lose the $3s$ electron to approach the stable inert gas structure of Neon so the ionization energy should drop when going from Ne to Na and we see that it does.

From here on, the chart exhibits similar breaks due to the reasons given above. We see that the noble gases have the highest ionization energies, which is what we would expect from our knowledge of their great stability and inactivity.

Problem Assignment

For each of the elements given below, diagram as in Problem No. 9-1, (a) the number of protons; (b) the number of neutrons; (c) the number of electrons in each energy level; and (d) the number of electrons in each of the atomic orbitals.

9-1.

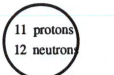

2) 8) 1)

$1s^2 \, 2s^2 \, 2p^6 \, 3s^1$

Na, atomic number 11,
mass number 23

9-2. C, atomic number 6,
mass number 13

9-3. Al, atomic number 13,
mass number 27

9-4. B, atomic number 5,
mass number 10

9-5. O, atomic number 8,
mass number 17

9-6. O, atomic number 8,
mass number 18

9-7. P, atomic number 15,
mass number 31

9-8. Mn, atomic number 25,
mass number 56

9-9. Fe, atomic number 26,
mass number 57

9-10. Fe, atomic number 26,
mass number 58

9-11. Si, atomic number 14,
mass number 29

9-12. S, atomic number 16,
mass number 32

9-13. Ar, atomic number 18,
mass number 38

9-14. Ne, atomic number 10,
mass number 20

9-15. Mg, atomic number 12,
mass number 24

9-16. Mg, atomic number 12,
mass number 26

9-17. V, atomic number 23,
mass number 51

9-18. Ni, atomic number 28,
mass number 59

9-19. Cl, atomic number 17,
mass number 35

9-20. K, atomic number 19,
mass number 41

Chapter 10

THE ELECTRON

1. *The Bohr Atom*

Introducing a volatile compound into a flame or passing a high voltage electric discharge through a gas will produce light. Dispersion of this light by passing it through a grating or quartz prism results in atomic spectra or line spectra. The lighter elements such as hydrogen yield a relatively small number of lines while the heavy elements may yield hundreds of lines. The spectral lines may lie in the infrared, the ultraviolet, and the X-ray ranges as well as in the visible (colored) range of electromagnetic radiation.

The Bohr theory explained the spectral lines in terms of an electron moving from one energy level to another of lower energy. With an electronic transition from an outer energy level to the first energy level, $n=1$ (also called the ground state), a series of lines called the Lyman series is produced. The Lyman series lies in the ultraviolet range. Electromagnetic radiation is emitted as photons of light when an electron descends from a higher quantum level to one of a lesser quantum number.

Energy must be introduced to cause an electron to ascend from a lower energy level to higher energy level. When an electron jumps from an outer energy level to the second energy level ($n=2$), the Balmer series of lines is produced. This series lies in the near ultraviolet and in the visible range. When an electron jumps from an outer energy level to the 3rd, 4th, or 5th energy level, the series are called Paschen, Brackett, and Pfund respectively. These three series lie in the infrared range.

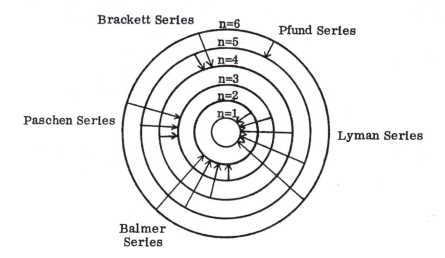

2. *Energy Relationships*

The energy emitted by an electron as it moves from one energy level to another of lower energy may be represented in terms of the frequency of the emitted light wave by the equation

$$E = h\nu$$

where E is the energy in ergs, h is Planck's constant (6.626 $\times$ 10^{-34} joule sec.) and ν is the frequency of the emitted wave.

The SI unit for frequency is the hertz (Hz). 1 Hz = 1 cycle/second and since cycles have no units, hertz has the units "per second" or 1/sec. Larger frequency units are the kilohertz and megahertz.

Then, since $c = \nu\lambda$ (velocity=frequency$\times$wavelength), $\nu=c/\lambda$. Substituting this value for ν in the above equation,

$$E = h\nu = h\frac{c}{\lambda} = hc \cdot \frac{1}{\lambda}$$

This relationship may also be written as

$$E = -E_0 Z^2 \left(\frac{1}{n_1^2} - \frac{1}{n_2^2} \right)$$

where n_1 is the energy level *from* which the electron is moving and n_2 is the energy level *to* which the electron is moving. E_0 is a constant equal to 2.179×10^{-18} joules and Z is the atomic number.

$n_2 = 1$ for the Lyman series

$n_2 = 2$ for the Balmer series

$n_2 = 3$ for the Paschen series

$n_2 = 4$ for the Brackett series

$n_2 = 5$ for the Pfund series

Example No. 10-1   An electron falls from the 5th energy level to the 2nd energy level in the hydrogen atom. Calculate the energy emitted in joules (J) and in electron volts.

Using the formula:

$$E = -E_0 Z^2 \left(\frac{1}{n_1^2} - \frac{1}{n_2^2} \right) \text{ where } Z = 1 \text{ for hydrogen}$$

$$E = -2.179 \times 10^{-18} \text{ J } (1)^2 \left(\frac{1}{5^2} - \frac{1}{2^2} \right)$$

$$E = 4.576 \times 10^{-19} \text{ J}$$

Next, since 1 electron volt (1 eV) = 1.602×10^{-19} J

$$4.576 \times 10^{-19} \text{ J} = \frac{4.576 \times 10^{-19} \text{ J}}{1.602 \times 10^{-19} \text{ J/eV}} = 2.856 \text{ eV}$$

Example No. 10-2.   A sample of hydrogen was bombarded with electrons. Radiation in the Lyman series was observed when the bombarding electrons had energies of 12.75 eV. From what energy level did the electrons drop?

Since the radiation is in the Lyman series, the electrons must have dropped to the first energy level ($n_2 = 1$).

Converting electron volts to joules,

12.75 eV = 12.75 eV $\times$ 1.602×10^{-19} J/eV = 2.043×10^{-18} J

Then using the formula $E = -E_0 Z^2 \left(\dfrac{1}{n_1^2} - \dfrac{1}{n_2^2} \right)$

$$2.043 \times 10^{-18} \text{ J} = -2.179 \times 10^{-18} \text{ J } (1)^2 \left(\frac{1}{n_1^2} - \frac{1}{n_2^2} \right) \quad \text{where } Z = 1$$

$$\frac{1}{n_1^2} - \frac{1}{1^2} = \frac{2.043 \times 10^{-18} \text{ J}}{-2.179 \times 10^{-18} \text{ J}} = -0.9380 \qquad (\text{where } m_2 = 1)$$

$$\frac{1}{n_1^2} = 1 - 0.9380 = 0.0620$$

$$n_1^2 = \frac{1}{0.0620} = 16 \text{ and } n_1 = \sqrt{16} = 4$$

Example No. 10-3. The wavelength of a spectral line of the hydrogen atom is 95.0 nm. What is the frequency of the light wave? What is its energy? If the spectral line is due to an electron dropping from the fifth energy level, into which series will the line fall?

Since $c = v\lambda$, $v = \dfrac{c}{\lambda}$ ($c = 3.00 \times 10^8$ m/sec; $h = 6.626 \times 10^{-34}$ joule sec.)

$$v = \frac{3.00 \times 10^8 \text{ m/sec.}}{95.0 \times 10^{-9} \text{ m}} = 3.16 \times 10^{15} \text{ cycles/sec.}$$

Then, $E = hv = 6.626 \times 10^{-34}$ joule sec. $\times 3.16 \times 10^{15}$ cycles/sec. $= 2.09 \times 10^{-18}$ J

Next, using $E = -E_0 Z^2 \left(\dfrac{1}{n_1^2} - \dfrac{1}{n_2^2} \right)$

$$2.09 \times 10^{-18} \text{ J} = -2.179 \times 10^{-18} \text{ J} \left(\frac{1}{5^2} - \frac{1}{n_2^2} \right) \quad \text{where } Z = 1$$

$$\frac{1}{25} - \frac{1}{n_2^2} = \frac{2.09 \times 10^{-18}}{-2.179 \times 10^{-18}} = -0.960$$

$$\frac{1}{n_2^2} = 0.960 + \frac{1}{25} = 1.000$$

$$n_2^2 = 1.000$$

$$n_2 = 1$$

Since the electron is dropping back to the first energy level ($n_2 = 1$), the spectral line will be in the **Lyman** series.

Example No. 10-4. What is the frequency of the light emitted when an electron falls from $n = 5$ to $n = 3$ in the hydrogen atom? Velocity of light $= 2.998 \times 10^8$ m/sec. What is the wavelength in nm?

$$E = -E_0 Z^2 \left(\frac{1}{n_1^2} - \frac{1}{n_2^2} \right) = -2.179 \times 10^{-18} \text{ J } (1)^2 \left(\frac{1}{5^2} - \frac{1}{3^2} \right)$$

$$E = 1.550 \times 10^{-19} \text{ J}$$

Next, recalling that $E = h\nu$, $\nu = \dfrac{E}{h}$

$$\nu = \frac{1.550 \times 10^{-19} \text{ J}}{6.626 \times 10^{-34} \text{ J sec.}} = 2.340 \times 10^{14} \text{ cycles/sec.} = 2.340 \times 10^{14} \text{ Hz.}$$

Finally, since $c = \nu\lambda$, $\lambda = \dfrac{c}{\nu} = \dfrac{2.998 \times 10^8 \text{ m/sec.}}{2.340 \times 10^{14} \text{ cycles/sec.}} = 1.281 \times 10^{-6} \text{ m}$

$$1.281 \times 10^{-6} \text{ m} \times \frac{10^9 \text{ nm}}{1 \text{ m}} = 1.281 \times 10^3 \text{ nm} = 1281 \text{ nm}$$

3. Size of the Bohr Atom

An electron moving in a circular path around a nucleus can have an angular momentum given by the formula

$$mvr = nh/2\pi \qquad\qquad 3\text{-A}$$

where m is the mass of the electron, v its velocity, r the radius of its orbit, n the quantum number of the orbit, and h Planck's constant.

This electron must also satisfy classical mechanics so that the centrifugal force on the electron (mv^2/r) is balanced by the centripetal force (the electrostatic force) (Ze^2/r^2), so that

$$mv^2/r = Ze^2/r^2 \qquad\qquad 3\text{-B}$$

Solving equation 3-A for r,

$$r = \frac{nh}{2\pi mv} \qquad\qquad 3\text{-C}$$

Solving equation 3-B for r (on the left side of the equation)

$$r = \frac{mv^2 r^2}{Ze^2} \qquad\qquad 3\text{-D}$$

Squaring r from equation 3-C and substituting for r^2 in 3-D,

$$r = \frac{mv^2}{Ze^2} \left(\frac{nh}{2\pi mv} \right)^2 = \frac{n^2 h^2}{4\pi^2 mZe^2}$$

Note that in this equation h, π, m, and e are constant. Also for a given element Z is constant so that r varies directly as n^2.

Example No. 10-5. Calculate the radius in cm for an electron in the 4th energy level of the hydrogen atom.

The equation $r = \dfrac{n^2h^2}{4\pi^2mZe^2}$ may be rewritten as $r = r_0n^2$ where

$$r_0 = h^2/4\pi^2mZe^2 = 5.29 \times 10^{-9} \text{cm (for hydrogen)}$$

So, $\qquad r = r_0n^2 = 5.29 \times 10^{-9}\text{cm}(4)^2 = 8.46 \times 10^{-8}\text{cm}.$

4. Wavelength of an Electron

An electron, like light, has properties of both a particle and a wave. The DeBroglie wavelength of a particle moving at a velocity small in comparison with that of light is given by the formula

$$\lambda = \frac{h}{mv}$$

For particles travelling at velocities near that of light, the Compton wavelength is given by formula

$$\lambda = \frac{h}{m_0c}\left(\frac{\sqrt{1 - v^2/c^2}}{v/c}\right) \text{ where } c \text{ is the velocity of light and } v \text{ is the velocity of the particle.}$$

For an electron, $h/m_0c = 2.426 \times 10^{-10}\text{cm}.$

Example No. 10-6. Calculate the wavelength of an electron travelling at 0.9 velocity of light.

$$\lambda = \frac{h}{m_0c}\left(\frac{\sqrt{1 - v^2/c^2}}{v/c}\right) = 2.426 \times 10^{-10} \text{ cm } \frac{\sqrt{1 - (0.9c/c)^2}}{0.9c/c}$$

$$= 2.426 \times 10^{-10} \frac{\sqrt{1 - 0.81}}{0.9} \text{ cm}$$

$$= 2.426 \times 10^{-10} \frac{\sqrt{0.19}}{0.9} \text{ cm}$$

$$= 1.175 \times 10^{-10} \text{ cm}$$

Example No. 10-7. Calculate the wavelength of an oxygen molecule moving at a velocity of 3.00×10^5 cm/sec.

We can use the formula $\lambda = \dfrac{h}{mv}$ but since h is in joule sec., m must be in kg and v in m/sec.

1 mole of oxygen weighs 32.0 g and contains 6.02×10^{23} molecules,

1 molecule of oxygen weighs $\dfrac{32.0}{6.02\times10^{23}}$ g or 5.32×10^{-23} g or 5.32×10^{-26} kg.

The velocity of the oxygen molecule is 3.00×10^5 cm/sec. or 3.00×10^3 m/sec.

Then, $\lambda = \dfrac{h}{mv} = \dfrac{6.626\times10^{-34} \text{ joule sec.}}{5.32\times10^{-26} \text{ kg}\times3.00\times10^3 \text{ m/sec.}} = 4.15 \times 10^{-12}$ m.

Checking the units, since joules are equivalent to kg $m^2/\text{sec.}^2$,

$$\lambda = \frac{h}{mv} = \frac{\text{joule sec.}}{\text{kg m/sec.}} = \frac{\text{kg m}^2/\text{sec.}^2 \times \text{sec.}}{\text{kg m/sec.}} = \text{m}$$

Problem Assignment

10-1.   Calculate the energy in joules produced when an electron descends from the 15th energy level to the 1st energy level in the hydrogen atom. .... $(2.169 \times 10^{-18}$ J)

10-2.   How many electron volts are required to ionize a hydrogen atom (make the electron ascend from $n=1$ to $n=$infinity)?

10-3.   A light wave in the Lyman series of Hydrogen emits radiation with an energy of 13.22 eV. From what energy level did the electron jump? .... (6)

10-4.   A wave in the Lyman series of Hydrogen emits radiation with an energy of 12.75 eV. From what energy level did the electron jump?

10-5.   What is the energy of light radiation with a frequency of 4.00×10^{18} cycles/sec.? What is the wavelength of the light wave? .... $(2.65 \times 10^{-15}$ J, 0.0750 nm)

10-6.   Light with a wavelength of 680.0 nm has what energy in electron volts? in Joules?

10-7.   Calculate the energy (in electron volts) required when an electron is moved from the 3rd energy level to the 6th energy level in the hydrogen atom. .... (1.13 eV)

10-8.   The wavelength of a line in the Brackett series of Hydrogen is 2630 nm. What is its frequency? What is its energy? From what energy level to what energy level is the electronic transition occurring?

10-9.   Calculate the frequency of the first line in the Paschen spectral series of the hydrogen atom. .... $(1.599 \times 10^{14}$ Hz)

10-10. A spectral line in the Balmer series of the hydrogen atom has a wavelength of 368.28 nm. From what energy level is the electron moving?

10-11. A spectral line in the Brackett series of the hydrogen atom has a wavelength of 4050 nm. From what energy level to what energy level is the electron moving? .... (5 to 4)

10-12. What is the wavelength of the light produced when an electron drops from infinity to the 4th energy level of the hydrogen atom?

10-13. What is frequency of the 5th line in the Lyman spectral series of Hydrogen? .... $(3.198 \times 10^{15}$ Hz)

10-14. The wavelength of a spectral line of hydrogen is 1.8751 nm. What is its energy? If the electron leaves the 4th energy level, into which energy level is it entering? .... $(1.059 \times 10^{-19}$ J, 3)

10-15. Which produces more energy, an electron moving from the 5th to the 4th energy level, or from the 6th to the 5th energy level in the hydrogen atom? What is the energy differential?

10-16. Calculate the radius of an electron in the 3rd energy level of the hydrogen atom. .... $(4.76 \times 10^{-8}$ cm)

10-17. In which energy level of the hydrogen atom will the electron's path have a radius of 3.39×10^{-7} cm?

10-18. Calculate the radius of an electron in the 3rd energy level of a hydrogen atom.

10-19. Calculate the radius for an electron in the 12th energy level of a hydrogen atom. .... $(7.62 \times 10^{-7}$ cm)

10-20. Calculate the DeBroglie wavelength of an electron moving at a velocity of 3.000 × 10^{6}cm/sec. . . . . (2.426 × 10^{-6} cm)

10-21. Calculate the DeBroglie wavelength of a proton moving with a velocity of 2.000 × 10^4 cm/sec. (mass of proton = 1.672 × 10^{-24} g).

10-22. Calculate the DeBroglie wavelength of a hydrogen molecule moving with a velocity of 3.000 × 10^5 cm/sec. . . . . (6.599 × 10^{-9} cm)

10-23. Calculate the DeBroglie wavelength of a neutron (rest mass 1.675 × 10^{-24} g) moving with a velocity of 1.000 × 10^6 cm/sec.

10-24. Calculate the wavelength of a electron travelling at 0.990 c.

10-25. Calculate the wavelength of a proton travelling at 0.995 c (mass of proton = 1.672 × 10^{-24} g).

Chapter 11

IONIC CHARGE AND OXIDATION NUMBER

The chemical activity of an element is due to its outermost electrons, those in its highest energy level. The electrons in the highest energy level are called the valence electrons. It is very tedious to draw the structure of an element with all of its electrons when we are concerned only with the outermost electrons, so we will use the electron dot formulas where the symbol of the element represents the kernel which consists of everything but the highest energy level. A dot represents one electron in the highest energy level. Thus, for Lithium, instead of Li 2) 1) we use Li• where the dot represents the one electron in the highest energy level. Calcium, Ca 2) 8) 8) 2) would be written as Ca•̣, where the 2 dots signify that Calcium has 2 electrons in its highest energy level. For Chlorine, instead of Cl 2) 8) 7) we have •• •• where the seven dots mean seven electrons in the highest energy level.

I. THE IONIC BOND

All elements tend to reach the rare gas structure by gaining, losing or sharing their outermost electrons. An atom which has lost or gained electrons is called an ion. If it has gained an electron or electrons, it forms a negative ion (since electrons are negatively charged); if it has lost an electron or electrons, it forms a positive ion.

So Sodium, Na•, will lose its 1 electron to give a positive Sodium ion, or Na^+ where the plus sign above the symbol indicates a positive charge.

Fluorine, •F•, will gain an electron to give a negative Fluoride ion, •F•⁻. We say Sodium has a charge of +1, meaning that it has lost 1 electron. Likewise, Fluorine has a charge of –1 since it has gained 1 electron. Thus, a positive ionic charge is the number of electrons lost by an atom, and a negative ionic charge is the number of electrons gained by an atom.

If we write the reaction between Na or Sodium and F or Fluorine, we have

Na • ⟍ + ⟶ •F• ⟶ Na^+ + •F•⁻

where the Sodium atom has lost its one outer electron to the Fluorine yielding a positive Sodium ion and a negative Fluoride ion. The positively charged Sodium ion and the negatively charged Fluoride ion would be expected to attract because of the attraction of opposite charges for one another. Therefore, the two ions would be held together by the electrostatic attraction of their charges. This is called an *ionic bond*. We can define it as follows: *an ionic bond results when there is a complete transfer of electrons from one atom to another, leaving positive and negative ions which attract one another by electrostatic forces.*

A few examples of ionic bonds should help in understanding this definition.

Mg •̣ ⟶ Cl, Cl ⟶ Mg^{2+} + •Cl•⁻ + •Cl•⁻

The Mg has lost 2 electrons, giving it a charge of +2. Each Chlorine has gained one electron, giving it a charge of –1. The compound $MgCl_2$ is held together by ionic bonds.

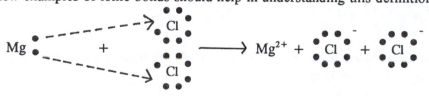

Ca •̣ - - - + - - → •O• ⟶ Ca^{2+} + •O•²⁻

The calcium has lost 2 electrons and has a charge of +2 while the oxygen which has gained the 2 electrons has a charge of −2. Here again we have an ionic bond between the Calcium and the Oxygen.

In the reaction Al• + •O•, we see that the aluminum atom has 3 outer electrons and the oxygen has 6. Another rule we should mention here is that metals always lose all their outer electrons and never only part of them. That is, the aluminum atom must lose all 3 of its outer electrons. However, the oxygen atom can gain only two of these electrons to fill its outer energy level to 8. The least common multiple for a loss of 3 electrons (from the aluminum) and a gain of 2 electrons (for the oxygen) is 6. The loss of electrons by the aluminum (3) goes into the least common multiple twice, so we need 2 aluminum atoms. The gain by the oxygen (2) goes into the least common multiple 3 times so we need 3 oxygens. Thus the equation becomes:

$$2 \text{ Al}\bullet + 3 \bullet\text{O}\bullet \rightarrow 2 \text{ Al}^{3+} + 3 \bullet\text{O}\bullet^{2-}$$

Each Aluminum has lost 3 electrons, giving it a charge of +3. Each Oxygen has gained 2 electrons, giving it a charge of −2. We note here that the number of electrons lost must always equal the number of electrons gained. Two Aluminums lost 3 electrons each or a total of 6 electrons while the three Oxygens gained 2 electrons each or a total of 6 electrons.

II. THE COVALENT BOND

There is another way in which atoms can complete their outer energy level—by sharing electrons. Consider the molecule H_2, formed from two hydrogen atoms, each with one outer (valence) electron.

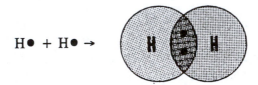

H• + H• →

The hydrogen atoms share electrons so that each has a completed outer energy level, 2 in this case, since the first energy level can hold only 2 electrons. Such a bond between the hydrogen atoms is called a covalent bond and may be represented either by a pair of shared electrons, as indicated above, or by H—H where the dash represents a covalent bond, a pair of shared electrons. Note how this type of bond differs from an ionic bond where there is a complete transfer of electrons.

Chlorine has 7 outer electrons, •Cl•. Two chlorine atoms may combine and share electrons to form a chlorine molecule, Cl_2, where each atom has 8 outer electrons.

•Cl• + •Cl• → Cl : Cl   or   •Cl — Cl•   or simply as Cl — Cl

In general, in order to attain greater stability, atoms tend towards a noble gas structure with 8 outer electrons (except for hydrogen which needs only 2 electrons).

For more complex molecules we use the following rules to predict the number and placement of shared electrons.

Rule 1. *Determine the number of outer electrons for each atom in the compound.*

Rule 2. *Calculate the total number of outer electrons for all of the atoms in Rule 1.*

Rule 3. *Find the number of electrons each atom needs in order to fill its outer energy level to 8, except for hydrogen which needs to fill its outer energy level to 2.*

Rule 4. *Calculate the total number of electrons for all atoms in Rule 3.*

Rule 5. *Divide the total number of electrons needed from Rule 4 by 2 to obtain the number of shared electron pairs in the given compound.* Note: if your answer is not a whole number, then such a group of atoms probably cannot reach a stable structure.#

Rule 6. *Draw a symmetrical structure for the atoms in the compound.* (Assume that all compounds are symmetrical unless otherwise indicated).

Rule 7. *Distribute the electron pairs (Covalent bonds) from step 5 between the atoms in the symmetrical structure, remembering that each hydrogen can form only one bond.*

Rule 8. *Determine the number of outer electrons not needed for the formation of bonds by subtracting the total number of electrons needed for bonding (from Rule 4) from the total number of outer electrons available (from Rule 2).*

Rule 9. *Fill in the electrons from Rule 8 onto the symmetrical structure so that each atom has an outer energy level of 8 (except for hydrogen which has only 2).*

Let us try a few examples to see how these rules may be used.

Consider the compound NH_3. What type of bonds are present and where are these bonds located?

First, since nitrogen is a nonmetal and also since hydrogen when written last in a formula acts as a nonmetal, there can be no ionic bonds present. An ionic bond exists between a metal and a non-metal. Therefore, covalent bonds must be present, so we use the preceding rules.

Rule 1. Determine the number of outer electrons for each atom in the compound.

In this compound, NH_3, we have one nitrogen and three hydrogens with the electron dot structures of

$$\bullet \, N \, \bullet \quad \text{and} \quad H\bullet$$

Rule 2. Calculate the total number of outer electrons for all of the atoms in Rule 1.

Number of outer electrons in nitrogen	= 5
Number of outer electrons in 3 hydrogens	= 3
Total number of outer electrons	= 8

Rule 3. Find the number of electrons each atom needs in order to fill its outer energy level to 8, except for hydrogen which needs only 2.

Number of electrons needed by nitrogen	= 3
Number of electrons needed by each hydrogen	= 1

Rule 4. Calculate the total number of electrons for all atoms in Rule 3.

Nitrogen needs	3 electrons
Three hydrogens need	3 electrons
Total number of electrons needed	6 electrons

Such compounds, with half pairs (unbound electrons) do exist but they will not be discussed in this book.

Rule 5. Divide the total number of electrons needed for Rule 4 by 2 to obtain the number of shared electron pairs in the given compound.

$$\frac{6}{2} = 3 \text{ electron pairs}$$

Rule 6. Draw a symmetrical structure for the atoms in the compound.

H

H   N   H

Rule 7. Distribute the electron pairs from step 5 in the symmetrical structure, remembering that a hydrogen can form only one bond (can share only one electron pair).

Rule 8. Determine the number of outer electrons not needed for the formation of bonds by subtracting the total number of electrons needed for bonding (Rule 4) from the total number of outer electrons available (Rule 2).

Total number of outer electrons (Rule 2)   = 8

Total number of electrons needed for bonding (Rule 4)   = 6

Number of outer electrons not needed in bonding   = 2

Rule 9. Fill in the electrons from Rule 8 so that each atom has an outer energy level of 8 except for hydrogen which has only 2.

Note that each hydrogen already has 2 outer electrons so that the remaining 2 electrons must be placed around the nitrogen, or

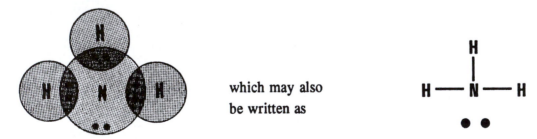

which may also
be written as

Thus in NH_3 there are 3 covalent bonds (pairs of shared electrons) and also one pair of unshared (unbonded) electrons.

Next let us consider the N_2 molecule.

Rule 1. Each nitrogen has 5 outer electrons, or :N: .

Rule 2. Total number of outer electrons = 10.

Rule 3. Number of electrons needed by each nitrogen to fill outer energy level to 8 = 3.

Rule 4. Total number of electrons needed to fill outer energy level of both nitrogens to 8 = 6.

Rule 5. $\frac{6}{2}$ = 3 shared electron pairs.

Rule 6.   N   N

Rule 7.   N $\overset{\bullet\ \bullet}{\underset{\bullet\ \bullet}{\bullet\ \bullet}}$ N

Rule 8. Number of outer electrons not needed for shared pairs = 10 – 6 = 4.

Rule 9.   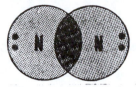   or

Since, by definition, a covalent bond is represented by a pair of shared electrons, in the case of N_2 above, we have a triple covalent bond present since there are three pairs of shared electrons between the nitrogens.

   <u>Now consider the compound CCl₄</u>.

Rule 1. $\bullet$ C $\bullet$ and $\overset{\bullet\bullet}{\underset{\bullet\bullet\bullet}{\bullet}}$ Cl $\bullet$

Rule 2. Carbon has 4 outer electrons.

   4 chlorines have a total of 28 outer electrons.

   Total number of outer electrons = 32.

Rule 3. Number of electrons needed by C to fill outer energy level to 8 = 4.

   Number of electrons needed by Cl to reach outer energy level of 8 = 1.

Rule 4. Total number of electrons to fill all outer energy levels to 8 = 8.

Rule 5. $\frac{8}{2}$ = 4 pairs of shared electrons or 4 covalent bonds.

Rule 6.   Cl
          Cl   C   Cl
             Cl

Rule 7.   Cl
          Cl $\overset{\bullet\bullet}{\underset{\bullet\bullet}{\bullet\ \ \bullet}}$ C $\overset{\bullet}{\underset{\bullet}{}}$ Cl
             Cl

Rule 8. Number of outer electrons not needed for shared pairs = 32 – 8 = 24.

Rule 9.

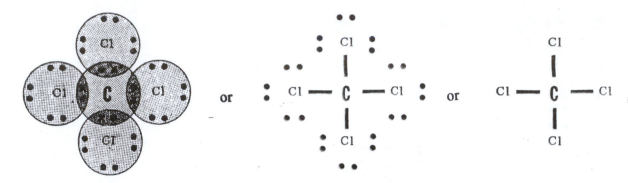

Next let us consider the compound CO_2. We will combine some of the rules in order to simplify the process.

Carbon has 4 outer electrons and each oxygen has 6, so that the total number of outer electrons is 16.

Carbon needs 4 more outer electrons and each oxygen needs 2 more outer electrons, so total number of outer electrons needed is 8.

Number of shared electron pairs is $\frac{8}{2}$ or 4, which means 4 pairs of shared electrons between the carbon and the oxygens. In order to have a symmetrical structure, we have

$$O \overset{\bullet\ \bullet}{\underset{\bullet\ \bullet}{\vdots}} C \overset{\bullet\ \bullet}{\underset{\bullet\ \bullet}{\vdots}} O$$

Since there are a total of 16 outer electrons and since 8 of them are used in bonding, there are 8 outer electrons left around the carbon and the oxygens, or

 or $$\overset{\bullet\bullet}{O} = C = \overset{\bullet\bullet}{\underset{\bullet\bullet}{O}}$$

Note that in this compound there are two double covalent bonds.

Now consider the sulfite ion, SO_3^{2-}. The charge of -2 indicates that the ion (a polyatomic ion) has gained 2 electrons from some other atom or atoms. In such cases we will assume that the gained electrons are added to the central atom, Sulfur in this case.

The sulfur atom has 6 outer electrons and each oxygen also has 6 outer electrons.

The sulfur atom needs 0 outer electrons since it originally had 6 and had gained 2 more from another source.

Each oxygen atom needs 2 outer electrons.

Total number of outer electrons needed for bonding = 6.

Number of shared electron pairs (covalent bonds) = 3.

Original number of outer electrons = 26 (6 from S, 6 from each 0, and 2 from other sources).

Number of outer electrons remaining after those used in bonding = 20.

Thus, the structure of SO_3^{2-} is

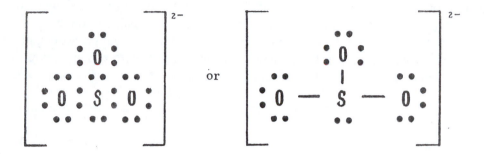

where there are 3 covalent bonds present.

There is still another method in which two atoms may be bonded together. This is the case where one atom donates all the electrons which are to be shared while the other donates none.

Look at the reaction: $NH_3 + H^+ \rightarrow NH_4^+$ or

The H^+ ion is covalently bonded to the Nitrogen by a shared electron pair but these electrons are supplied by the Nitrogen only so that the bond between the N and the H^+ is a coordinate covalent bond.

III. ELECTRONEGATIVITY

We note from the above that some atoms have a tendency to lose electrons while others have a tendency to gain electrons. A measure of this tendency is called the electronegativity of the atom. Thus an atom with a high electronegativity will be one which has a great attraction for electrons while an atom with a low electronegativity will have a small attraction for electrons. From this definition we see that metals will have a low electronegativity since they do not tend to gain but rather to lose electrons easily and non-metals will have a high electronegativity since they do tend to gain electrons.

Let us take an example of two elements in Group VII, Fluorine, F, and Iodine, I. Both of these are non-metals and have 7 outer electrons. In the case of F, highest energy level is much closer to the nucleus than it is in the case of I. Therefore, the positive nucleus of F will have a greater attractive force for electrons than will the nucleus of I since the F, being smaller, can get much closer to the transferable electron.

Since Fluorine, F 2)7 is the smallest atom with 7 outer electrons, it is the most electronegative of all elements. However, next to F 2)7 comes O 2)6 instead of Cl 2) 8) 7. This is because the O atom, while having 6 outer electrons instead of 7 as in Cl, is actually smaller in size and, therefore, is able to attract electrons more readily. Following is a table of the electronegativities of the common elements starting with F which is arbitrarily assigned an electronegativity of 4.0 and going down through the active metals.

F	4.0	Te, H, P	2.1	Y, Sc	1.3
O	3.5	As, B	2.0	Mg	1.2
N, Cl	3.0	Si, Sb	1.8	Sr, Li, Ca	1.0
Br	2.8	Sn, Ge	1.7	Na, Ba	0.9
C, S	2.5	Zr, Ti	1.6	K, Rb	0.8
I, Se	2.4	Al, Be	1.5	Cs	0.7

An atom of high electronegativity (a strong non-metal) will react with an atom of low electronegativity (a strong metal) to form an ionic bond. If no atom of low electronegativity is available, atoms of high electronegativity will share electrons with each other to form covalent bonds.

Let us draw the structures of a few simple covalently bonded compounds and see how the electronegativity will determine the position of the electrons. In the case of F_2, we have the two very electronegative atoms sharing electrons. Since the atoms are identical, the electrons must be shared equally. In the case of H-F, we again have a covalent bond between the two atoms but this time they are not of equal electronegativity. The F with the higher electronegativity will have a greater attraction for the shared electrons than H with its lower electronegativity. Therefore, even though this is a covalent bond and the electrons are shared between the H

and the F, they will be located closer to the F than to the H or H $\overset{\bullet\bullet}{\underset{\bullet\bullet}{\bullet F \bullet}}$. In other covalently bonded

compounds we can approximate the position of the shared electrons by comparing the electronegativities of the two atoms forming the bond.

IV.   OXIDATION NUMBER

In a compound containing ionic bonds, the oxidation number of an atom (or group of atoms) existing as an ion is the same as the charge of that ion. Thus, in the ionic compound $MgCl_2$, the oxidation number of the magnesium is +2 and that of the chlorine -1. In compounds having covalent bonds, we assign oxidation numbers according to the following rules:

1. Oxygen always has an oxidation number of -2 in any compound except in peroxides where it is -1.

2. The algebraic sum of all oxidation numbers in any compound must equal zero.

3. The algebraic sum of all the oxidation numbers in any polyatomic ion must equal the charge on the polyatomic ion.

4. All metals except those which are in a transition series have only one oxidation number.

A few examples should clarify these rules.

Example No. 11-1. $KMnO_4$ or $K^+ MnO_4^-$. Each Oxygen has an oxidation number of -2 so that 4 of them will total -8. K is not a transition metal and has only 1 oxidation number which is +1. Therefore, since the sum must equal zero and we have -8 for the Oxygens and +1 for the K, the Mn must have an oxidation number of +7 or

$$+1+7-8=0$$
$$K\ MnO_4$$

Example No. 11-2. H_3PO_3. Three Oxygens will have a total oxidation number of -6; three Hydrogens, since each is +1, will total +3. Therefore, in order to get a sum of zero, the P must have an oxidation number of +3.

$$+3+3-6=0$$
$$H_3\ P\ O_3$$

Example No. 11-3. $KClO_2$. Two Oxygens will have a total oxidation number of -4, one K will have an oxidation of +1 so that the Cl will have an oxidation number of +3, or

$$+1+3-4=0$$
$$K\ Cl\ O_2$$

Example No. 11-4. $(NO_3)^-$. This is a polyatomic ion so that according to the above rules the sum of the oxidation numbers must equal -1. Three Oxygens equal -6, and, therefore, the N, in order to give a total of -1, must have an oxidation number of +5.

$$+5-6=-1$$
$$N\ O_3$$

Example No. 11-5. BaO_2 or Barium peroxide. According to rule no. 1 above, the oxidation number of each Oxygen in a peroxide is -1 so that the total in this case for the two Oxygens is -2 leaving the Ba a $+2$. It is necessary to know whether a compound is a peroxide or a dioxide. For example, Na_2O_2 is sodium peroxide while SiO_2 is Silicon dioxide which is not a peroxide. Laboratory experience should tell us which compounds are peroxides and which are not.

V. WRITING FORMULAS FROM KNOWN OXIDATION NUMBERS

The writing of formulas when the oxidation numbers of the constituent ions are known is really very simple. All we have to remember is that THE TOTAL OF THE POSITIVE CHARGES MUST EQUAL THE TOTAL OF THE NEGATIVE CHARGES.

Example No. 11-6. Write the formula for Calcium Chloride.

From the following table, we see that Ca has an oxidation number or charge of $+2$ and Cl a -1. To equalize 2 positive charges we need 2 negative charges so we take 2 Cl's. Thus we have $Ca^{2+}Cl_2^{2-}$ or $CaCl_2$ (with charges understood but not written).

Example No. 11-7. Complete the formula for magnesium phosphate $Mg(PO_4)$.

Mg has an oxidation number of $+2$ and PO_4 an oxidation number of -3 so we have $Mg^{2+} PO_4^{3-}$ (to be completed).

The lowest common multiple between 2 and 3 is 6 so there must be 6 positive charges and 6 negative charges. Each Mg has a $+2$ charge so it takes 3 of them to make a total of $+6$. Each PO_4 has a charge of -3 so it takes 2 of them to make -6. Thus we have $Mg_3(PO_4)_2$.

TABLE OF COMMON OXIDATION NUMBERS

+1	+2	+3	+4	-1	-2	-3
Na	Ca	Al	Sn (ic)	Cl	O	PO_4
K	Mg	Sc		Br	S	BO_3
Li	Sr	Fe (ic)		NO_3	Cr_2O_7	N
H	Ba	Bi		OH	SO_4	
NH_4	Zn	Sb		CN	SO_3	
Cu (ous)	Cu (ic)	Cr		ClO_4	CrO_4	
Ag	Fe (ous)			MnO_4	CO_3	
Rb	Cd			ClO_3		
Cs	Pb			ClO_2		
	Sn (ous)			ClO		
	Be			NO_2		
	Hg (ic)			I		
	Ni			HCO_3		
	Co			$C_2H_3O_2$ (acetate)		
	Hg_2 (ous)			F		

Give the electron dot structures for the following elements:

11-1. Mg atomic number 12

11-2. Ti atomic number 22

11-3. S atomic number 16

11-4. Na atomic number 11

11-5. Ar atomic number 18

11-6. B atomic number 5

11-7. Si atomic number 14

11-8. Cl atomic number 17

11-9. Zn atomic number 30

11-10. Draw complete diagrams for the formation of compounds from the following atoms (assuming ionic bonds):
a) Cs & Cl   b) Mg & F   c) Al & S

Draw the electron dot structures for the following compounds, labelling the various types of bonds present.

11-11. H_2O	11-12. BaF_2	11-13. $NaCl$	11-14. ZnO
11-15. H_2S	11-16. PH_3	11-17. $CHBr_3$	11-18. PCl_3
11-19. KBr			

What is the oxidation number of the underlined atom or ion?

11-20. $\underline{Rb}_2S$	11-21. $Mg\underline{Cl}_2$	11-22. $Na\underline{N}O_3$	11-23. $\underline{P}H_3$
11-24. $H_2\underline{SO}_3$	11-25. $\underline{Fe}Cl_2$	11-26. $Mg(\underline{ClO}_3)_2$	11-27. $\underline{Si}O_2$
11-28. $K_2\underline{Cr}_2O_7$	11-29. $Na_2\underline{S}O_3$	11-30. $\underline{Cu}Cl_2$	11-31. $Na\underline{Bi}O_3$
11-32. $K\underline{Mn}O_4$	11-33. $H_2\underline{Mn}O_4$	11-34. $Mg_3\underline{N}_2$	11-35. $Ag_3\underline{P}O_4$
11-36. $\underline{B}Cl_3$	11-37. $\underline{Pt}Cl_4$	11-38. $H_3\underline{P}O_2$	11-39. $Al_2(\underline{SO}_4)_3$
11-40. $\underline{Sr}CO_3$	11-41. $\underline{Co}SO_4$	11-42. $\underline{Fe}(OH)_3$	11-43. $H_2\underline{Se}$
11-44. $Ca\underline{Si}O_3$	11-45. $K_2\underline{Mn}O_4$	11-46. $K_3\underline{As}O_3$	11-47. $\underline{As}_2O_3$
11-48. $MgNH_4\underline{P}O_4$	11-49. $NH_4\underline{N}O_2$	11-50. $\underline{Sc}_2O_3$	11-51. $\underline{Cs}_2O$
11-52. $Na_2\underline{Zn}O_2$	11-53. $Zn(H\underline{C}O_3)_2$	11-54. $K\underline{Br}O_2$	11-55. $\underline{N}_2H_4$
11-56. $\underline{Ni}Cl_2$	11-57. $K_2\underline{Pt}Cl_6$	11-58. $Na_2\underline{Sn}(OH)_4$	11-59. $\underline{Cu}Cl$
11-60. $\underline{Ag}_2O$	11-61. $Mg\underline{C}_2O_4$	11-62. $Al\underline{N}$	11-68. $K\underline{I}O_4$
11-64. $\underline{Os}O_4$	11-65. $\underline{V}_2O_5$	11-66. $H\underline{Au}Cl_4$	11-67. $\underline{Ti}O_2$
11-68. $\underline{Cr}_2(SO_4)_3$	11-69. $\underline{Pb}O_2$	11-70. $\underline{Cd}S$	11-71. $\underline{Cl}_2O_7$
11-72. $\underline{W}O_3$	11-73. $\underline{C}_6H_{12}O_6$	11-74. $\underline{Hg}_2Cl_2$	11-75. $\underline{Mn}_2O_3$
11-76. $H_3\underline{As}O_3$	11-77. $\underline{Sb}_2S_3$	11-78. $H_2\underline{Te}$	11-79. $\underline{Y}_2O_3$

Complete the following formulas by adding the proper subscripts wherever necessary.

11-80. H S	11-81. Ca SO_4	11-82. Sr OH	11-83. Ca N
11-84. Mg PO_4	11-85. Al PO_3	11-86. Pb NO_3	11-87. Li SO_4
11-88. H PO_4	11-89. NH_4 CN	11-90. K CrO_4	11-91. Rb CO_3
11-92. Li S	11-93. Ca ClO	11-94. Ag PO_4	11-95. Al OH
11-96. Mg CN	11-97. Sr NO_2	11-98. NH_4 PO_4	11-99. H NO_2
11-100. Cd NO_3	11-101. Na SO_3	11-102. Ag O	11-103. H CO_3
11-104. NH_4 NO_2	11-105. Zn CO_3	11-106. Sn Cl (ous)	11-107. Cs MnO_4
11-108. Na SO_3	11-109. Ag Cr_2O_7	11-110. Al ClO_4	11-111. Al MnO_4
11-112. Mg ClO_3	11-113. Ca MnO_4	11-114. Fe SO_4 (ic)	11-115. Rb O
11-116. Ca H	11-117. Al O	11-118. Mg N	11-119. Cs S
11-120. K HCO_3	11-121. Zn I	11-122. Pb O	11-123. Na SO_4
11-124. Ba PO_4	11-125. Cu S (ous)	11-126. Li CO_3	11-127. Rb Br
11-128. Ba O (peroxide)	11-129. K O (peroxide)		

11-130. Complete the following table with the proper formulas:

	Chloride Cl^-	Hydroxide OH^-	Nitrate NO_3^-	Sulfate SO_4^{2-}	Sulfide S^{2-}	Carbonate CO_3^{2-}	Phosphate PO_4^{3-}
Hydrogen H^+							
Sodium Na^+							
Ammonium NH_4^+							
Potassium K^+							
Calcium Ca^{2+}							
Magnesium Mg^{2+}							
Ferrous Fe^{2+}							
Stannous Sn^{2+}							
Aluminum Al^{3+}							
Ferric Fe^{3+}							
Stannic Sn^{4+}							

Chapter 12

ATOMIC AND MOLECULAR ORBITALS

I. ATOMIC ORBITALS

The $1s$ orbital, also called the $1s$ atomic orbital, represents a spherical electron probability or electron cloud about the atomic nucleus. This spherical electron cloud represents the space which has the highest probability of containing the electron. That is, the probability of finding the $1s$ electron within that spherical electron cloud is nearly 100%. The shape of the $1s$ atomic orbital (the $1s$ AO) and its arrangement along the X, Y, and Z axes may be represented as:

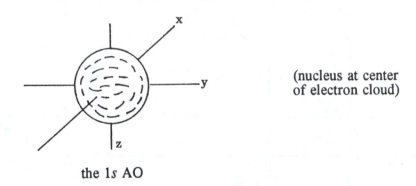

(nucleus at center of electron cloud)

the $1s$ AO

The $2s$ and $3s$ atomic orbitals are also spherical. However, as the number of the energy level increases, the radius of the electron cloud also increases. Thus the $2s$ AO is larger in volume than the $1s$ AO, and the $3s$ AO is even larger than the $2s$ AO. That is, there is a greater probability of finding the $3s$ electron farther from its nucleus than the $2s$ electron and there is a greater probability of finding the $2s$ electron farther from the nucleus than the $1s$ electron.

In each energy level except the first, there is one s AO and three p AO's. These three p atomic orbitals are arranged spatially along the X, Y, and Z axes and are designated as p_x, p_y, and p_z respectively.

Since there are three p orbitals, each capable of holding a maximum of 2 electrons, the three p orbitals can hold a maximum of 6 electrons. However, the sequence of filling the p orbitals is such that each of the orbitals, the p_x, p_y, and p_z, must hold one electron before a second electron can fill any of these orbitals. Thus the sequence of filling the $2p$ AO's is:

(1) $2p_x^1$

(2) $2p_x^1 \, 2p_y^1$

(3) $2p_x^1 \, 2p_y^1 \, 2p_z^1$

(4) $2p_x^2 \, 2p_y^1 \, 2p_z^1$

(5) $2p_x^2 \, 2p_y^2 \, 2p_z^1$

(6) $2p_x^2 \, 2p_y^2 \, 2p_z^2$

A p atomic orbital has a slightly distorted dumbbell shape with one lobe on either side of the nucleus. The p AO's represent the maximum electron density in a given direction or the highest probability of finding the p electron in the space about the nucleus.

The p_x AO may be represented spatially as:

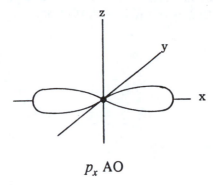

p_x AO

Likewise the p_y and p_z AO's may be represented as:

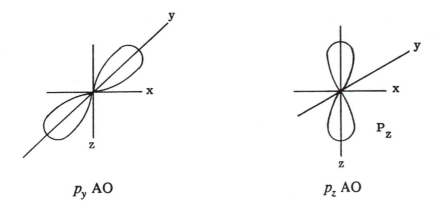

p_y AO p_z AO

All three p AO's may be represented together about the nucleus as:

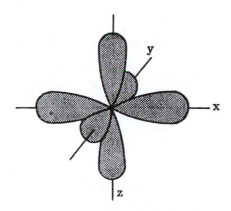

Spatial arrangement of the $3p$ AO's

Since there are five d orbitals in an energy level, there must be five d atomic orbitals in that energy level. Four of these d AO's have a cloverleaf shape. The fifth d AO is similar in shape to a p AO but with a donut-shaped ring around its nucleus. The shapes of the five d AO's are:

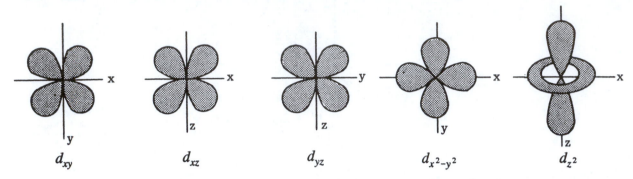

d_{xy} $\qquad$ d_{xz} $\qquad$ d_{yz} $\qquad$ $d_{x^2-y^2}$ $\qquad$ d_{z^2}

Example No. 12-1. Diagram the atomic orbitals for hydrogen, atomic number 1.

Hydrogen has only one electron which goes into the s AO of the first energy level, or $1s^1$. The s orbital is spherical in shape, so the diagram of the hydrogen atom is:

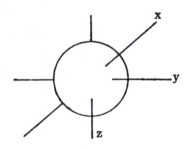

Hydrogen, $1s^1$

with the nucleus being at the center of the electron cloud.

Example No. 12-2. Diagram the outermost electron cloud for the element boron, atomic number 5.

The electron configuration for boron, atomic number 5, is:

$1s^2\ 2s^2\ 2p^1$, or when written spatially,

$1s^2\ 2s^2\ 2p_x^{\,1}$.

The p_x atomic orbital may be diagrammed as:

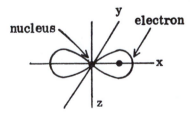

with a dot representing the one electron in the p_x AO.

Example No. 12-3. Diagram the p AO's for nitrogen, atomic number 7.

The electron configuration for nitrogen, atomic number 7, is:

$1s^2\ 2s^2\ 2p^3$.

Written in terms of the x, y, and z axes, the arrangement is:

$$1s^2\ 2s^2\ 2p_x^1\ 2p_y^1\ 2p_z^1$$

The structure of the p AO's for nitrogen is:

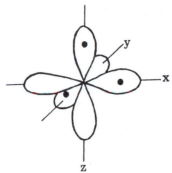

As before, the nucleus is at the center of the electron cloud. As with most AO diagrams, the inner filled AO's are omitted for the sake of clarity. In this diagram each electron is designated by a dot in the appropriate AO.

Example No. 12-4. Diagram the outer AO's for chlorine, atomic number 17.

The electron configuration for chlorine is:

$$1s^2\ 2s^2\ 2p^6\ 3s^2\ 3p^5,\ \text{or}$$

$$1s^2\ 2s^2\ 2p_x^2\ 2p_y^2\ 2p_z^2\ 3s^2\ 3p_x^2\ 3p_y^2\ 3p_z^1$$

We will indicate filled atomic orbitals (those containing 2 electrons) by means of shading, so that the structure for chlorine's $3p$ AO's is:

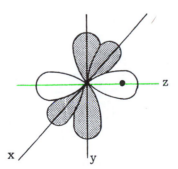

As before, the nucleus is at the center of the electron cloud and the inner filled atomic orbitals are omitted. The unpaired electron in the p_z AO is indicated by a dot.

II. BONDING OF ATOMIC ORBITALS

The electron-dot structure for the H_2 molecule as shown at the right indicates that two hydrogen atoms have combined to form a hydrogen molecule with a covalent bond between them. If we consider the two hydrogen atoms, designated arbitrarily as H_a and H_b, we can diagram their respective $1s$ atomic orbitals as:

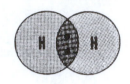

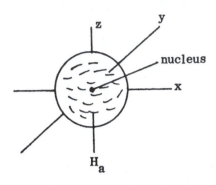

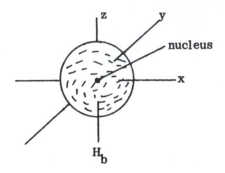

When these two hydrogen atoms combine to form a hydrogen molecule, their atomic orbitals overlap as indicated in the following diagram:

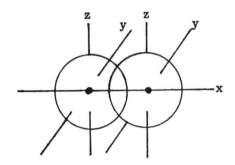

overlapping AO's of two H atoms

and then form the following type of electron cloud:

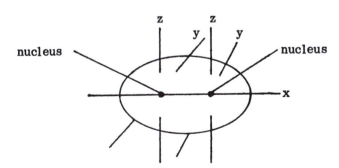

This combination of two s atomic orbitals forms a larger electron cloud which contains 2 electrons.

The bond formed by the overlapping of the two s atomic orbitals of the hydrogen atoms is called a sigma (σ) bond and like all sigma bonds it is cylindrically symmetrical. A sigma bond is a covalent bond resulting from an increased electron density along the internuclear axis.

Example No. 12-5. Diagram the bond formed in the F_2 molecule.

The electron configuration for fluorine, atomic number 9, is:

$$1s^2\ 2s^2\ 2p^5 \text{ or spatially, } 1s^2\ 2s^2\ 2p_x^2\ 2p_y^2\ 2p_z^1$$

This may be diagrammed as:

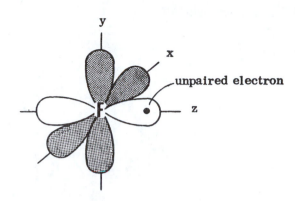

where the shaded atomic orbitals contain 2 (paired) electrons and so are unavailable for bonding. The p_z atomic orbital contains only 1 electron and so is available for bonding. (The inner, filled $1s$ and $2s$ AO's are omitted for the sake of clarity.)

The two fluorine atoms comprising the fluorine molecule, here arbitrarily designated as F_a and F_b, can be diagrammed as:

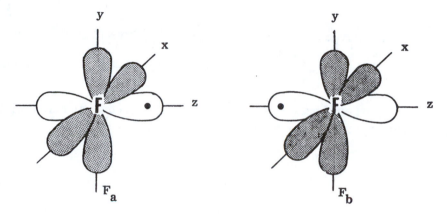

When these p_z AO's overlap, they may do so as illustrated below

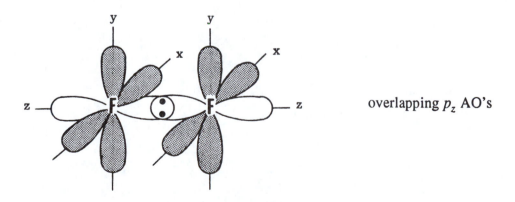

overlapping p_z AO's

The combined p_z AO's may be diagrammed as:

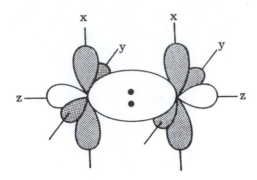

This type of bond is a sigma bond because it is the result of an overlapping of the p_z AO's along the internuclear axis and hence is cylindrically symmetrical. The electron dot structure for F_2 is

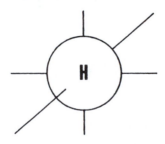 so that we can see that a sigma bond corresponds to a covalent bond.

Example No. 12-6. Diagram the type of bond formed in the $HCl_{(g)}$ molecule.

The electron-dot structure for $HCl_{(g)}$ is H Cl with a covalent bond between the H and the Cl.

Therefore, there should be a sigma bond between the H and Cl atoms when the structure is diagrammed. The structure for hydrogen, atomic number 1, electron configuration $1s^1$ is:

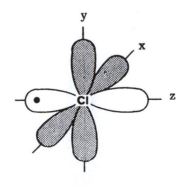

The structure for chlorine, atomic number 17, electron configuration

$$1s^2 \; 2s^2 \; 2p_x^2 \; 2p_y^2 \; 2p_z^2 \; 3s^2 \; 3p_x^2 \; 3p_y^2 \; 3p_z^1 \text{ is:}$$

with the shaded atomic orbitals being filled (contained paired electrons). Thus the atomic orbitals available for bonding are the $1s$ in the hydrogen and the $3p_z$ in the chlorine.

These available atomic orbitals may be designated as:

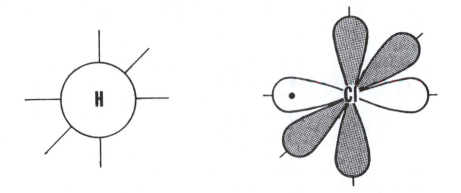

When these atomic orbitals overlap, they form the following type of electron cloud.

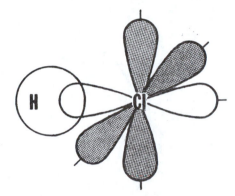

Thus overlap of the $1s$ AO from the hydrogen atom with the $3p_z$ AO from the chlorine atom produces a sigma bond because we can see that there is an increased electron density along the internuclear axis. Note that a sigma bond may be formed by the overlap of one s AO with another s AO, (as in H_2), by the overlap of an s AO with a p AO (as in HCl), and by the overlap of a p AO with another p AO (as in F_2).

Example No. 12-7. Diagram the structure of the H_2S molecule.

The sulfur atom, atomic number 16, has the electron configuration of

$$1s^2\ 2s^2\ 2p_x^2\ 2p_y^2\ 2p_z^2\ 3s^2\ 3p_x^2\ 3p_y^1\ 3p_z^1$$

and the hydrogen atom, atomic number 1,
$$1s^1$$

The atomic orbitals available for bonding in the sulfur atom are the $3p_y$ and $3p_z$ while the $1s$ is available in the hydrogen atom.

(Recall that atomic orbitals which contain 2 electrons are filled and are unavailable for bonding.)

The AO's available for bonding in the sulfur atom may be diagrammed as follows:

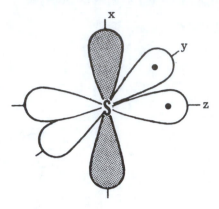

with the shaded AO being filled and unavailable for bonding. These are the $3p$ AO's. The inner AO's are all filled, unavailable for bonding, and so are not diagrammed here for the sake of clarity.

The p_y and p_z AO's may each overlap with a $1s$ AO of a hydrogen atom to form the following type of structure:

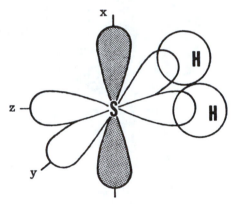

with a sigma bond formed between each hydrogen and the sulfur atom.

Since the p_y and p_z AO's are arranged along the Y and Z axes, and since these axes are at right angles with each other, the bond angle between the hydrogens should be 90°. The actual bond angle is 92°, which is in close agreement with the predicted value.

On the basis of the predicted bond angles in the H_2S molecule, the electron-dot structure for this molecule is:

with a covalent bond between each of the hydrogens and the sulfur atom.

Pi Bonds

If two *p* AO's overlap, as shown below:

or

a sigma bond results. Note that there is an increased electron density along the internuclear axis.

However, if two *p* AO's overlap sideways, as shown below:

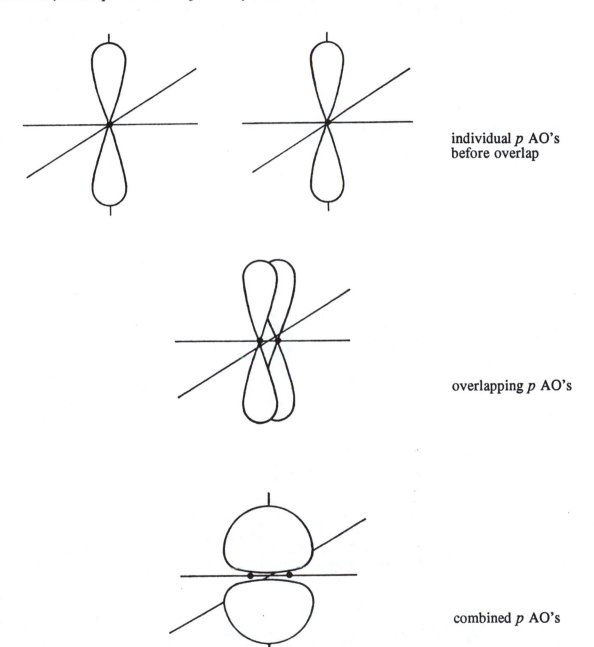

individual *p* AO's
before overlap

overlapping *p* AO's

combined *p* AO's

a pi (π) bond results. A *pi bond* is a covalent bond resulting from the overlap of *p* AO's in a side by side fashion producing an increased electron density above and below the internuclear axis.

Example No. 12-8. Diagram the structure of the S_2 molecule.

The electron configuration for sulfur, atomic number 16, is

$$1s^2\ 2s^2\ 2p_x^2\ 2p_y^2\ 2p_z^2\ 3s^2\ 3p_x^2\ 3p_y^1\ 3p_z^1$$

with the $3p_y$ and $3p_z$ AO's available for bonding.

If the $3p_y$ AO of one sulfur atom overlaps with the $3p_y$ AO of a second sulfur atom, the following structure will result:

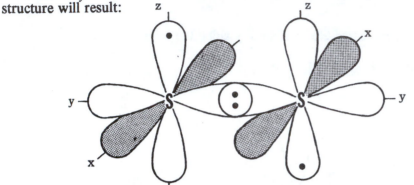

(Note order of axes used for clarity of presentation)

with a sigma bond being formed between the two sulfur atoms. However, we can see that the p_z AO's are also available for bonding. If the p_z AO of one sulfur atom overlaps the p_z AO of the other sulfur atom (a sideways overlap), then the following structure will be formed:

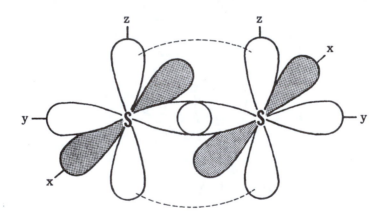

which will become:

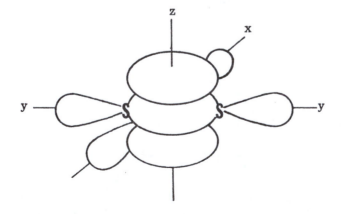

Thus the sideways overlap of the p_z AO's produces a pi bond. Therefore, in the S_2 molecule there is a sigma and a pi bond between the sulfur atoms. The electron-dot structure for S_2 is:

so that a double covalent bond consists of one sigma bond and one pi bond.

Example No. 12-9. Diagram the structure of the nitrogen molecule, N_2.

The nitrogen atom, atomic number 7, has the electron configuration

$$1s^2\ 2s^2\ 2p_x^{\ 1}\ 2p_y^{\ 1}\ 2p_z^{\ 1}$$

and the structure

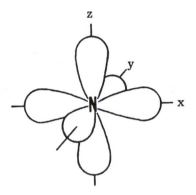

If two nitrogen atoms have an overlap of their p_x AO's, then a sigma bond will be formed between these atoms, or

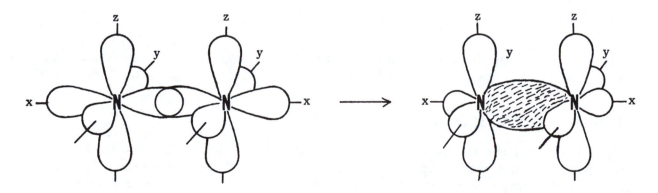

If the two p_y AO's overlap sideways, a pi bond will be formed, or

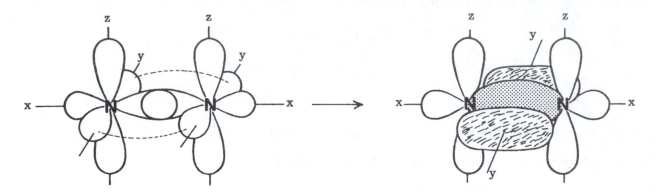

If the two p_z AO's overlap sideways another pi bond will be formed. Thus the structure of the N_2 molecule is:

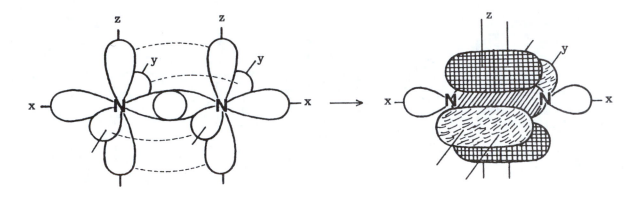

The electron-dot structure of the N_2 molecule is:

$$: N :: : : N :$$

with a triple covalent bond between the nitrogen atoms. That is, a triple covalent bond consists of a sigma bond and two pi bonds. (Recall that a double covalent bond consists of a sigma bond and one pi bond.)

III. HYBRID ATOMIC ORBITALS

Hybridization is the process of combining two or more atomic orbitals within one atom for satisfactory covalent bonding of that atom to other atoms. The net effect of hybridization is to change an atom with electrons in lower energy atomic orbitals to an atom with electrons relocated into effective bonding atomic orbitals with slightly higher energy.

Let us see how this works. Consider the beryllium atom, atomic number 4. The electron configuration for Be is $1s^2 2s^2$. From this electron arrangement, both s atomic orbitals are full and so this element should be relatively inert. However, beryllium does form the covalent compound BeH_2 so that there must be some bonding atomic orbitals available in the beryllium atom. (Recall that bonding atomic orbitals can contain only one electron each).

If one of the $2s$ electrons is promoted to a $2p$ atomic orbital (since the $2s$ and $2p$ AO's are energetically very close to one another), then the new electron configuration becomes $1s^2 2s^1 2p_x^1$ with the $2s$ and $2p_x$ AO's available for bonding. The $2s$ and $2p$ AO's can combine to form two mixed or hybrid AO's. This type of hybrid orbital is designated as an *sp* hybrid after the two AO's which formed it. The *sp* hybrid orbitals may be considered to have formed as indicated in the following diagram:

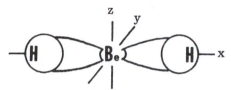

s AO p_x AO   sp hybrid   sp hybrid

Since each of these two *sp* hybrid orbitals contains one electron and since electrons repel one another, these two hybrid orbitals should be as far apart as possible, or at an angle of 180°.

Example No. 12-10. Diagram the structure of the BeH_2 molecule, using *sp* hybrid orbitals.

Each of the *sp* hybrid orbitals can bond with the $1s$ AO of a hydrogen atom to form the following structure.

In this compound there are two sigma bonds, one between the Be atom and each H atom. Since *sp* hybrid orbitals are 180° apart, the compound BeH_2 is linear.

Next consider the element boron, atomic number 5. The electron configuration for B is $1s^2\ 2s^2\ 2p_x^{\ 1}$ so that only the $2p_x$ AO appears to be available for bonding. However, boron forms the covalent compound BF_3 so that there must be three (equal) AO's available for bonding. If one of the $2s$ electrons is promoted to a $2p$ AO, then the electron configuration becomes $1s^2\ 2s^2\ 2p_x^{\ 1}\ 2p_y^{\ 1}$.

When an *s* AO combines with two *p* AO's, three sp^2 hybrid orbitals are formed. Each of these three sp^2 hybrid orbitals contains one electron and is available for bonding. The combination of one *s* AO with two *p* AO's to form sp^2 hybrid AO's is indicated in the following diagram:

s AO p_x AO p_y AO   3 sp^2 hybrid AO's

The three sp^2 hybrid orbitals all lie in one plane so that compounds having this type of structure are planar. Since the three hybrid orbitals are equivalent and equal, the angle between them is 120°.

Example No. 12-11. Diagram the structure of the BF_3 molecule, using sp^2 hybridization.

The F atom has the electron configuration $1s^2\ 2s^2\ 2p_x^2\ 2p_y^2\ 2p_z^1$ so that the $2p_z$ AO is available for bonding. The structure of fluorine may be represented as:

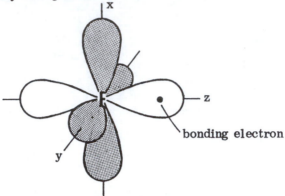

When the p_z AO of the F overlaps one of the sp^2 hybrid orbitals of the B atom, a sigma bond is formed. The three sigma bonds formed by the overlap of the three p_z AO's from the three F atoms with the three sp^2 hybrid orbitals of the B atom may be shown as:

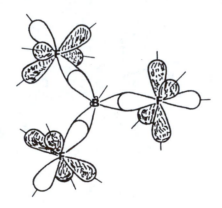

Now let us consider the element carbon, atomic number 6. The electron configuration for C is $1s^2\ 2s^2\ 2p_x^1\ 2p_y^1$ so that only two atomic orbitals appear to be available for bonding. However, in the compound CH_4 there are four equal covalent bonds so that there must be four atomic orbitals available for bonding. If one electron from the $2s$ AO is promoted to a $2p$ AO, then the electron configuration would be:

$$1s\ 2s^1\ 2p_x^1\ 2p_y^1\ 2p_z^1$$

The *s* AO can combine with three *p* AO's to form four hybrid orbitals which are designated as sp^3 hybrids. The four sp^3 hybrid orbitals form a figure known as a tetrahedron, as shown in the following diagram:

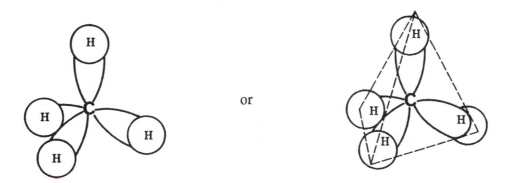

| *s* AO | three *p* AO's | four sp^3 hybrid orbitals |

The angle between the sp^3 hybrid orbitals is 109° 28'.

Example No. 12-12. Diagram the structure of the CH_4 molecule, using sp^3 hybrid orbitals.

Each hydrogen atom has a 1*s* AO available for bonding and the carbon atom has four sp^3 hybrid orbitals available, as shown in the preceding diagram. The structure of the compound CH_4 is:

or

Each hydrogen atom forms a sigma bond with one of the sp^3 hybrid orbitals of the carbon. The shape of the compound CH_4 is tetrahedral with a bond angle (the angle H–C–H) of 109° 28'.

The electron-dot structure for CH_4 is:

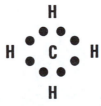

where again the (single) covalent bonds are equivalent to sigma bonds.

Combining the information about the various types of hybrid orbitals, we have:

Type of hybrid	Shape of molecule	Angle of bonds
sp	linear	180°
sp²	planar	120°
sp³	tetrahedral	109° 28'

Example No. 12-13. The compound C_2H_6 has the following electron-dot structure:

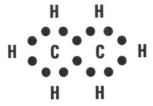

and is known to contain two tetrahedral carbon atoms. Diagram its structure.

The electron-dot structure indicates single bonds (covalent) between each hydrogen and a carbon atom and also between carbon atoms. These single covalent bonds must represent sigma bonds. The tetrahedral arrangement of the carbon atoms indicates sp^3 hybrid orbitals.

Therefore, overlapping one sp^3 hybrid orbital of one carbon atom with an sp^3 hybrid of the other carbon atom to have a sigma bond between them, we have:

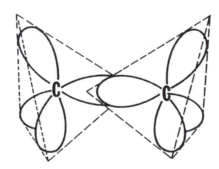

Overlapping each of the remaining sp^3 hybrid orbitals with the $1s$ AO of a hydrogen atom to give sigma bonds, we have the structure of C_2H_6 as:

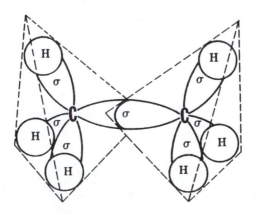

Example No. 12-14. The compound C_2H_4 is known to be a planar compound. Its electron-dot structure is:

Diagram its structure.

The single covalent bonds between the carbon and hydrogen atoms indicate sigma bonds. The double covalent bond between carbon atoms indicates a sigma and a pi bond. The planar structure indicates sp^2 hybrid orbitals. The electron configuration of carbon, atomic number 6, is

$$1s^2\, 2s^2\, 2p_x^{\,1}\, 2p_y^{\,1}$$

If we promote one of the $2s$ electrons to a $2p$ orbital, the configuration becomes

$$1s^2\, 2s^1\, 2p_x^{\,1}\, 2p_y^{\,1}\, 2p_z^{\,1}$$

If we need sp^2 hybrid orbitals, we must use one s AO and two p AO's leaving one p AO non-hybridized. This may be diagrammed as:

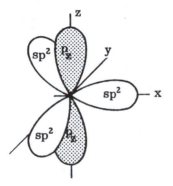

where the shaded AO is the non-hybridized one. Note that the sp^2 hybrid orbitals are all planar and at an angle of 120°.

If we allow one sp^2 hybrid orbital of each carbon to overlap, forming a sigma bond between them, we have:

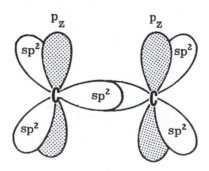

If each of the unused sp^2 hybrid orbitals overlap the $1s$ AO of a hydrogen atom the result is:

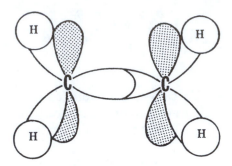

Now, if each unused p AO of the carbon atom overlaps sideways with the corresponding unused (and non-hybridized) AO of the other carbon atom to form a pi bond, we have the structure of C_2H_4 as:

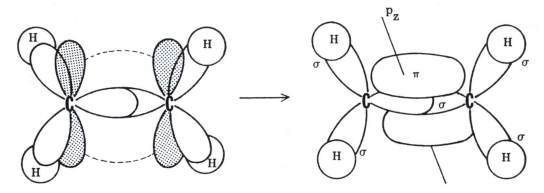

Example No. 12-15. The compound C_2H_2 is a linear compound with the following electron-dot structure. Diagram its orbital arrangement.

$$H \overset{\bullet}{\underset{\bullet}{\,}} C \overset{\bullet\,\bullet\,\bullet}{\underset{\bullet\,\bullet\,\bullet}{\,}} C \overset{\bullet}{\underset{\bullet}{\,}} H$$

Since this molecule is linear, we should expect it to contain sp hybrid orbitals. The single covalent bonds between the carbons and the hydrogens represent sigma bonds. The triple covalent bond between the carbons should represent a sigma bond and two pi bonds.

The electron arrangement of the carbon atom is

$$1s^2\, 2s^2\, 2p_x^{\,1}\, 2p_y^{\,1}$$

and after promoting one $2s$ electron to a $2p$ orbital it becomes

$$1s^2\, 2s^1\, 2p_x^{\,1}\, 2p_y^{\,1}\, 2p_z^{\,1}$$

If the 2*s* AO combines with one 2*p* AO to form two *sp* hybrid orbitals, the 2*p* AO is left non-hybridized, as shown in the following diagram:

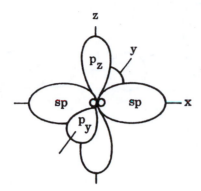

then two *sp* hybrid orbitals, one from each carbon atom, can overlap to produce a sigma bond, or

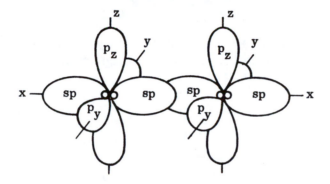

The other *sp* hybrid orbitals can each overlap with the 1*s* orbitals of a hydrogen atom, or

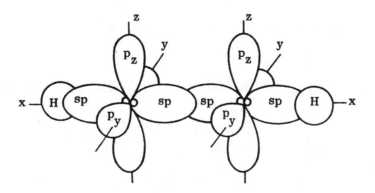

There are $2p$ AO's remaining on each of the carbon atoms. These can overlap sideways in pairs to form two pi bonds, or

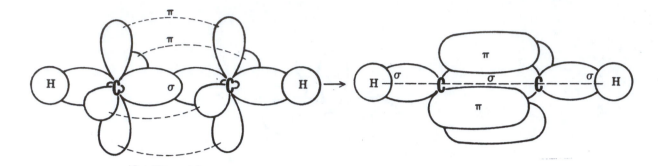

Example No. 12-16. The CO_2 molecule is known to be linear. Its electron-dot structure is given below. Diagram its orbital structure.

Since the CO_2 molecule is linear, the carbon atom must have sp hybrid orbitals. As indicated in the previous example, the sp hybrid orbitals for the carbon atom may be diagrammed as

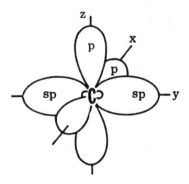

The oxygen atom, atomic number 8, has the electron configuration of

$$1s^2 \ 2s^2 \ 2p_x^2 \ 2p_y^1 \ 2p_z^1$$

If we diagram the p AO's of the oxygen atom, shading the p_x AO which is filled and unavailable for bonding, we have

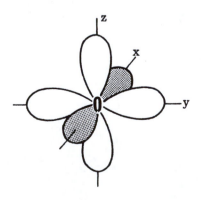

Now, if an *sp* hybrid orbital from the carbon atom overlaps the p_y AO from the oxygen atom we have

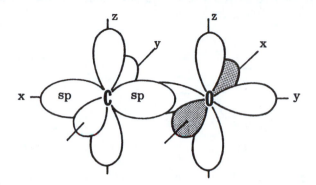

and if the other *sp* hybrid orbital from the carbon atom overlaps the p_y AO from the other oxygen atom,

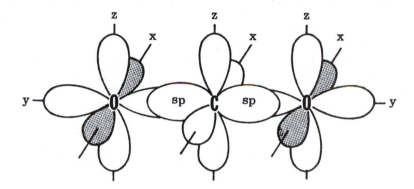

Finally, if the p_z AO's overlap sideways with an available p AO from the carbon atom, the structure of CO_2 becomes:

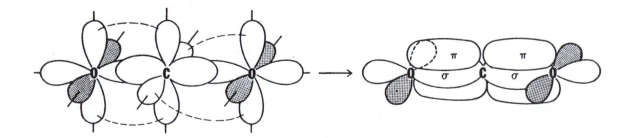

IV. MOLECULAR ORBITALS

In a molecular orbital the electrons are interacting with more than one nucleus. That is, a molecular orbital (MO) is characteristic of the molecule as a whole. For simple diatomic molecules we can combine atomic orbitals to form molecular orbitals by using the following rule: *The number of molecular orbitals formed is exactly equal to the total number of atomic orbitals combined.*

That is, when the 1s AO of a hydrogen atom reacts with the 1s AO of another hydrogen atom, two different MO's are possible. If the two 1s AO's react with an increased electron density between the nuclei, as shown below:

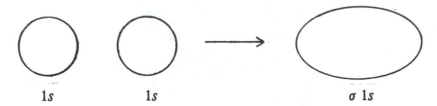

sigma bond or sigma bonding molecular orbital is produced. Such a sigma bonding molecular orbital, produced from 1s AO's is abbreviated as σ1s.

However, if the two 1s AO's combine in such a way that the electron density between the nuclei decreases, a sigma anti-bonding molecular orbital, or σ*1s is formed, where the * indicates an antibonding molecular orbital.

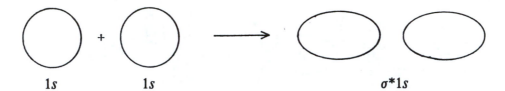

The stability of the 1s AO's, the σ 1s MO and the σ* 1s MO are indicated below:

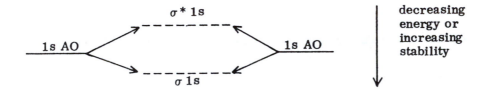

Thus we can see that the σ 1s MO is more stable than either of the 1s AO's which in turn are more stable than the σ* 1s MO.

Likewise, when two 2s AO's combine, σ 2s and σ* 2s MO's are formed with the σ 2s being more stable.

Using the preceding rule, three 2p AO's can form a total of six different MO's. If 2p AO's react along the internuclear axis to increase the electron density between the nuclei, then a σ 2p MO may be formed. If these p AO's react in such a manner that the electron density between the nuclei is decreased, then a σ* 2p MO is formed.

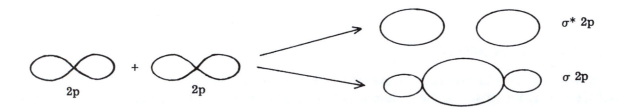

If two $2p$ AO's react above and below the internuclear axis, $\pi\ 2p$ and $\pi^*\ 2p$ MO's are possible.

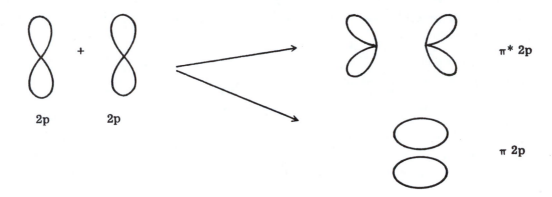

The order of stability of these bonding and anti-bonding molecular orbitals is:

σ_{1s} is more stable than σ^*_{1s}

σ^*_{1s} is more stable than σ_{2s}

σ_{2s} is more stable than σ^*_{2s}

σ^*_{2s} is more stable than σ_{2p_x}

σ_{2p_x} is more stable than π_{2p_y} and π_{2p_z} (of equal stability)

π_{2p_y} and π_{2p_z} are more stable than $\pi^*_{2p_y}$ and $\pi^*_{2p_z}$ (of equal stability)

$\pi^*_{2p_y}$ and $\pi^*_{2p_z}$ are more stable than $\sigma^*_{2p_x}$

Thus the molecular orbitals, in order of their stability, are:

$$\sigma^*_{2p_x}$$

$$\pi^*_{2p_y} \qquad\qquad \pi^*_{2p_z}$$

$$\pi_{2p_y} \qquad\qquad \pi_{2p_z}$$

$$\sigma_{2p_x}$$

$$\sigma^*_{2s}$$

$$\sigma_{2s}$$

$$\sigma^*_{1s}$$

$$\sigma_{1s}$$

The rules for filling these molecular orbitals are:

1. Each MO can hold a maximum of 2 electrons

2. Electrons fill the lowest available molecular orbitals. Higher energy MO's begin to fill only when lower energy MO's hold 2 electrons.

3. When two MO's have equal energy, the electrons go into each to give half-filled MO's before a second electron can enter to produce electron pairs.

Example No. 12-17. Indicate the MO's used in the O_2 molecule.

Since oxygen, atomic number 8, has a total of 8 electrons, the O_2 molecular has total of 16 electrons to be placed into various MO's. If we indicate the electrons by means of arrows, and indicate their opposing spins by having the arrows pointing in opposite directions, then the first 2 electrons can fill the σ_{1s} MO (see following diagram, part a).

The next 2 electrons fill the next energy level, the σ^*_{1s} (figure b). The fifth and sixth electrons fill the σ_{2s} MO (figure c) and the seventh and eighth electrons the σ^*_{2s} MO (figure d).

Orbital	(a)	(b)	(c)	(d)	(e)	(f)	(g)	(h)
$\sigma^*_{2p_x}$	—	—	—	—	—	—	—	—
$\pi^*_{2p_y}$ / $\pi^*_{2p_z}$	— —	— —	— —	— —	— —	— —	— —	↑ ↑
π_{2p_y} / π_{2p_z}	— —	— —	— —	— —	— —	↑ ↑	↑↓ ↑↓	↑↓ ↑↓
σ_{2p_x}	—	—	—	—	↑↓	↑↓	↑↓	↑↓
σ^*_{2s}	—	—	—	↑↓	↑↓	↑↓	↑↓	↑↓
σ_{2s}	—	—	↑↓	↑↓	↑↓	↑↓	↑↓	↑↓
σ^*_{1s}	—	↑↓	↑↓	↑↓	↑↓	↑↓	↑↓	↑↓
σ_{1s}	↑↓	↑↓	↑↓	↑↓	↑↓	↑↓	↑↓	↑↓

Then come the electrons in the σ_{2p_x} (figure e). Since the next two MO's have equal energy, one electron must go into each one first before a second electron can go into either. Thus the next two electrons go into these MO's as indicated in figure f. The following 2 electrons can fill these MO's as shown in figure g. The last 2 electrons go into the next higher energy level. Since there are two energy levels of equal energy, one electron goes into each as shown in figure h.

Thus there are 2 unpaired electrons in the oxygen molecule. This is confirmed by the fact that the oxygen molecule exhibits paramagnetism (due to unpaired electrons).

Thus the molecular orbital representation for the O_2 molecule is

$$\left(\sigma_{1s}\right)^2 \left(\sigma^*_{1s}\right)^2 \left(\sigma_{2s}\right)^2 \left(\sigma^*_{2s}\right)^2 \left(\sigma_{2p_x}\right)^2 \left(\pi_{2p_y}\right)^2 \left(\pi_{2p_z}\right)^2 \left(\pi^*_{2p_y}\right)^1 \left(\pi^*_{2p_z}\right)^1$$ where the superscripts indicate the number of electrons in each molecular orbital.

Problem Assignment

12-1.   Compare the size and shape of the $1s$ and $2s$ AO's

12-2.   How do the p AO's differ from one another?

12-3.   Diagram the outermost atomic orbitals for the following elements: carbon, atomic number 6; oxygen, atomic number 8; sodium, atomic number 11.

12-4.   Indicate the orbital arrangement of electrons, using x, y, and z spatial designations for the p AO's for: aluminum, atomic number 13; phosphorus, atomic number 15; argon, atomic number 18; lithium, atomic number 3.

12-5.   Which element is designated by the following AO's?

   a. $1s^2\ 2s^2\ 2p_x^{\ 1}$

   b. $1s^2\ 2s^2\ 2p_x^{\ 2}\ 2p_y^{\ 2}\ 2p_z^{\ 2}\ 3s^2\ 3p_x^{\ 2}\ 3p_y^{\ 2}\ 3p_z^{\ 2}\ 4s^1$

   c. $1s^2\ 2s^2\ 2p_x^{\ 2}\ 2p_y^{\ 2}\ 2p_z^{\ 2}\ 3s^2\ 3p_x^{\ 2}\ 3p_y^{\ 2}\ 3p_z^{\ 2}\ 4s^2\ 3d_{xy}^{\ 1}\ 3d_{xz}^{\ 1}$

12-6.   When two s AO's overlap, what type of bond is formed?

12-7.   When two p AO's overlap, what type of bonds are formed? Under what type of overlap does each occur?

12-8.   A single covalent bond corresponds to what type of bond involving atomic orbitals? A double covalent bond?

12-9.   Diagram the bond formed in the Cl_2 molecule showing the overlap of the AO's.

12-10. Diagram the structure of the HBr molecule.

12-11. Diagram the structure of the $P_{2(g)}$ molecule.

12-12. What is the bond angle in sp hybridization? sp^2? sp^3?

12-13. Diagram the structure of the tetrahedral CCl_4 molecule.

12-14. Diagram the structure of the planar compound C_2Cl_4.

12-15. Diagram the structure of the linear compound C_2Cl_2.

12-16. Which has a higher energy level, a bonding or an anti-bonding orbital?

12-17. Diagram the shaped of the σ_{1s} MO; σ^*_{1s} MO; π_{2p}; π^*_{2p}.

12-18. Diagram the molecular orbital arrangement for N_2.

12-19. Indicate the MO arrangement for Cl_2; for NO

12-20. What accounts for paramagnetism in a molecule?

Chapter 13

CRYSTAL SYSTEMS

Crystals can be classified into several systems. These crystal systems in turn may be further classified into sub-types.

A crystal is composed of many tiny unit cells. Each unit cell is the smallest part of the crystal which may be repeated over and over in three dimensions of space to form that crystal. The unit cell is defined in space by the lengths of its sides (a, b, and c) in each of the three dimensions and also by the angles (α, β, γ) at which the three planes meet.*

The following table indicates the different crystal systems.

Crystal System	Sides on Unit Cell	Angles of Unit Cell
Cubic	$a=b=c$	$\alpha=\beta=\gamma=90°$
Tetragonal	$a=b{*}c$	$\alpha=\beta=\gamma=90°$
Orthorhombic	$a{*}b{*}c$	$\alpha=\beta=\gamma=90°$
Hexagonal	$a=b{*}c$	$\alpha=\beta=90°$; $\gamma=120°$
Monoclinic	$a{*}b{*}c$	$\alpha=\gamma=90°$; $\beta{*}90°$
Triclinic	$a{*}b{*}c$	$\alpha{*}\beta{*}\gamma{*}90°$
Rhombohedral	$a=b=c$	$\alpha=\beta=\gamma{*}90°$

Example No. 13-1. A magnesium crystal is composed of unit cells with the following axes and angles:

$a=319$ pm, $b=319$ pm, $c=520$ pm
$\alpha=90°$, $\beta=90°$, $\gamma=120°$.

What types of crystal does magnesium form?

Since $a=b{*}c$ and since $\alpha=\beta=90°$ with $\gamma=120°$, the crystal type, according to the previous chart, must be hexagonal.

I. THE SIMPLE CUBE

Consider a unit cell which consists of a simple cube as shown in the following diagram:

Assume that there is an atom (sphere) at each corner of the cube and also assume that these spheres are in contact with one another.

* Length of sides of crystals are frequently measured in picometers (1 pm = 10-12m).

Example No. 13-2. How many complete spheres are present in the volume of each unit cell of the above simple cube?

Let us consider one corner of the cube. The sphere in that corner extends only part way into the volume of the unit cell we are considering. It also extends into other unit cells and is shared by them. How many unit cells of this shape (cube) can meet at a point? The answer is that there are eight unit cells meeting at a point, four in the plane above that point and four in the plane below. Therefore, only 1/8th of each corner sphere belongs to the unit cell under consideration. However, since the unit cell has 8 corners, there is an equivalent of 8 × 1/8 or one complete sphere per unit cell.

Example no. 13-3. What percent of the simple cube is empty space?

Let us assume that the edge of the simple cube is "a" units long. The total volume of that cube is a^3.

The radius of the spheres making up the cube (from the previous drawing) is seen to be $a/2$. The volume of the one sphere present in the simple cube can be calculated from the formula $Vol=4/3\pi r^3$ so that the volume of the sphere in the cube is $4/3 \times \pi \times (a/2)^3$ or $\pi a^3/6$.

Thus, the actual amount of space occupied in the simple cube is $\pi a^3/6$ while the total amount of space present is a^3.

Therefore, the amount of empty space is $a^3 - \pi a^3/6$ and % empty space =
$$\frac{a^3 - \pi a^3/6}{a^3} \times 100 = a^3 \left(\frac{1 - \pi/6}{a^3} \right) \times 100 = 47.7\%.$$

Example No. 13-4. Sodium chloride forms a simple cube with sodium ions and chloride ions alternating at the corners of the unit cell. Given the density of NaCl as 2.165 g/cm^3 and the molecular mass as 58.443, calculate (a) the size of the unit cell and (b) the radius of the sodium ion if the chloride ion has a radius of 181.2 pm.

Let us assume that each side of the unit NaCl cell is "a" units long. The volume of this unit cell is a^3. The mass of this unit cell can be calculated from its density and its volume.

$M = D \times V = 2.165$ g/cm^3 $\times a^3 = 2.165\, a^3$ g/cm^3.

Each unit cell of NaCl has 1/8th of each of its corner ions contributing to that cell. Each corner ion contributes only 1/8th of itself to the unit cell so that this unit cell really contains 1/2 Na$^+$ and 1/2 Cl$^-$ ions or 1/2 NaCl molecule.

We know that one molar mass of any compound contains Avogadro's number of molecules so that

58.443 g NaCl contain 6.023 × 10^{23} molecules of NaCl

1 molecule of NaCl has a mass of $\dfrac{58.443 \text{ g}}{6.023 \times 10^{23}}$

1/2 molecule of NaCl has a mass of $\dfrac{58.443 \text{ g} \times 1/2}{6.023 \times 10^{23}}$ or 4.851 × 10^{-23} g.

This is the mass of ½ molecule of NaCl and is also the mass of one unit cell of NaCl because that unit cell contains only ½ molecule of NaCl. So,

$2.165 \ a^3 \ g/cm^3 = 4.851 \times 10^{-23} \ g$

$$a^3 = \frac{4.851 \times 10^{-23} \ g}{2.165 \ g/cm^3} = 2.236 \times 10^{-23} \ cm^3$$

$$a = \sqrt[3]{2.236 \times 10^{-23} \ cm^3} = 2.817 \times 10^{-8} \ cm = 281.7 \ pm$$

Thus the edge of the unit NaCl cell is 281.7 pm long. This edge consists of a sodium ion and a chloride ion. Since the radius of the chloride ion is 181.2 pm, the radius of the sodium ion must be 281.7 pm – 181.2 pm, or 100.5 pm

Example No. 13–5. The density of a certain solid of molecular mass 58.100 is 2.480 g/cm³. This solid forms a simple cubic structure whose edges are 338.6 pm long. From this data, calculate Avogadro's number.

Knowing the density and the molecular mass of the solid, we can calculate the volume occupied by one mole of that solid,

$$V = \frac{M}{D} = \frac{58.100 \ g}{2.480 \ g/cm^3} = 23.43 \ cm^3$$

In a "molar" cube there must be Avogadro's number of unit cube cells. Each unit cell has an edge of 338.6 pm or 3386 × 10⁻⁸ cm. The volume of one of these unit cells is $(3.386 \times 10^{-8} \ cm)^3$ or $3.883 \times 10^{-23} \ cm^3$.

Then, the number of unit cells in the "molar" cube is Avogadro's number or

$$\text{Avogadro's number} = \frac{23.43 \ cm^3}{3.883 \times 10^{-23} \ cm^3} = 6.035 \times 10^{23}$$

Example No. 13–6. LiCl has a simple cubic structure. However, some of the lithium ion and an equal number of the chloride ion sites are vacant. The lattice is said to have some defects. If the length of the LiCl unit cell is 273 pm and if the observed density of solid LiCl is 1.70 g/cm³, calculate (a) the theoretical density of LiCl and (b) the percent of sites that are vacant.

The volume of the LiCl unit cell is $(273 \ pm)^3$ or $(2.73 \times 10^{-8} \ cm)^3$ or $2.035 \times 10^{-23} \ cm^3$.

As was mentioned for NaCl, each unit cell of LiCl in a simple cubic structure contains ½ molecule of LiCl (1/2 of an ion of Li^+ and 1/2 of an ion of Cl^-).

Therefore,

½ molecule of LiCl occupies $2.035 \times 10^{-23} \ cm^3$

1 molecule of LiCl occupies $4.070 \times 10^{-23} \ cm^3$ and

6.023×10^{23} molecules of LiCl occupy $4.070 \times 10^{-23} \ cm^3 \times 6.023 \times 10^{23}$ or $24.5 \ cm^3$.

Thus, the theoretical density of LiCl is its molecular mass divided by its molecular volume or

$$\text{Density (theoretical)} = \frac{42.4 \ g}{24.5 \ cm^3} = 1.73 \ g/cm^3$$

The given density of LiCl is 1.70 g/cm³.

Therefore, the percent of unoccupied sites is

$$\frac{1.73 \ g/cm^3 - 1.70 \ g/cm^3}{1.73 \ g/cm^3} \times 100 = 1.73\%.$$

II. THE BODY-CENTERED CUBE

Consider a cubic unit cell which has an atom at each corner and also an atom at its center. Such a cube is called a body-centered cube.

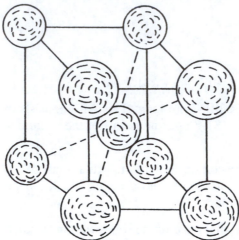

Example No. 13-7. How many atoms are present in each unit cell of a body-centered cube?

Just as in the simple cube, each corner atom of the body-centered cube contributes one-eighth of itself to that cube. Since there are eight corner atoms in a cube, the total contributions of the eight corner atoms is actually one atom. In addition to this atom contributed by all the corner atoms, there is also one atom in the center of the cube. This central atom belongs wholly to that cube. Therefore, in one unit cell of a body-centered cube there is an equivalent of two atoms present.

Example No. 13-8. Assuming that the corner atoms of a body-centered cube are in direct contact with the central atom, what percent of the body-centered cube is empty space?

Look at the above diagrammed body-centered cube. The corner atoms are not in direct contact with one another so that the edge "a" of this cube is not the same as twice the radius "r" of the corner atoms.

However, note that along the diagonal of the cube since we assumed that the corner atoms are in direct contact with the central atom, there are three atoms in direct contact. The length of this diagonal is $4r$ ($1r$ from the corner atom, $2r$ from the central atom, and $1r$ from the opposite corner atom).

The relationship between the diagonal of the cube, one edge of the cube, and the diagonal of one face of the cube can be drawn as:

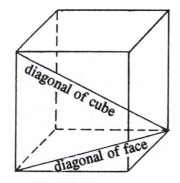

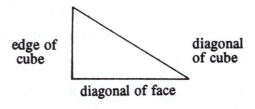

The length of the diagonal of one face of the cube can be calculated from the length "a" of its sides.

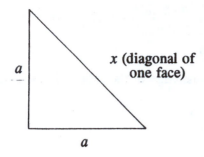

$$X^2 = a^2 + a^2 = 2\,a^2$$

$$X = a\sqrt{2}$$

Thus the diagram becomes

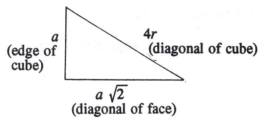

so, $(4\,r)^2 = (a)^2 + (a\sqrt{2})^2$

$16\,r^2 = a^2 + 2\,a^2 = 3\,a^2$

$a^2 = 16/3\ r^2$

$a = 4\sqrt{3}\ r/3$

The total volume of the cube is $(a)^3$ and since $a = 4\dfrac{\sqrt{3}\,r}{3}$,

total volume of cube is $\left[\dfrac{4\sqrt{3}\,r}{3}\right]^3$ or $64\dfrac{\sqrt{3}\,r^3}{9}$

The total volume of the 2 (equivalent) atoms present in the cube is

$$2\left[\frac{4}{3}\,\pi\,r^3\right] \text{ or } \frac{8}{3}\,\pi\,r^3$$

The amount of empty space in the cube is $\dfrac{64\sqrt{3}\,r^3}{9} - \dfrac{8}{3}\,\pi\,r^3$ and the percent of empty space =

$$\frac{\dfrac{64\sqrt{3}\,r^3}{9} - \dfrac{8}{3}\,\pi\,r^3}{\dfrac{64\sqrt{3}\,r^3}{9}} \times 100 = 32.1\%$$

Note that if the atoms are not in direct contact, then the percent of empty space will be greater because there will be more room between atoms.

Example No. 13-9.   Molybdenum crystallizes as a body-centered cube with a density of 10.2 g/cm³. Calculate the length of the edge of each unit cell of Mo. Assuming that the corner atoms of the unit cell are in contact with the central atom, calculate the atomic radius of molybdenum.

Each unit cell of Mo contains an equivalent of 2 atoms of Mo since it forms a body-centered cube. Therefore, in 1 molar mass there are 3.01×10^{23} unit cells.

The volume occupied by 1 molar mass of Mo (95.9 g) = $\dfrac{95.9 \text{ g}}{10.2 \text{ g/cm}^3}$ = 9.40 cm³.

Thus, 3.01 × 10²³ unit cells occupy 9.40 cm³ and 1 unit cell occupies $\dfrac{9.40 \text{ cm}^3}{3.01 \times 10^{23}}$ = 3.12 × 10⁻²³ cm³.

The edge of a unit cell = $\sqrt[3]{3.12 \times 10^{-23} \text{ cm}^3}$ = 3.15×10⁻⁸ cm = 315 pm

A triangle may be drawn consisting of the diagonal of the cube, one edge of the cube, and the diagonal of one face as shown below:

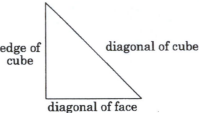

edge of cube | diagonal of cube

diagonal of face

The diagonal of the cube (three MO atoms in a line) = 4 r, where r is the atomic radius of Mo.

The length of the diagonal of the face of the cube = $a\sqrt{2}$ (where a is the length of the edge of the cube).

The above diagram thus becomes:

a | 4r | $a\sqrt{2}$

Therefore,
$$a^2 + (a\sqrt{2})^2 = (4\ r)^2$$
$$a^2 + 2\,a^2 = 16\ r^2$$
$$3\ a^2 = 16\ r^2$$
$$r^2 = 3/16\ a^2$$
$$r = \sqrt{3/16\ a^2} = \frac{a\sqrt{3}}{4} = \frac{315\ \text{pm}\sqrt{3}}{4} = 138\ \text{pm}$$
(since edge of the unit cell, a, = 315 pm)

Example No. 13-10. Chromium crystallizes in a body-centered cubic arrangement. If the atomic radius of chromium is 127 pm, calculate (a) the edge of the unit Cr cell and (b) the density of Cr.

Using the relationship and diagram of the previous example,

edge = a | 4r = diagonal of cube | $a\sqrt{2}$ = diagonal of face

$$a^2 + (a\sqrt{2})^2 = (4\ r)^2$$
$$3\ a^2 = 16\ r^2$$
$$a^2 = 16/3\ r^2$$
$$a = \frac{4\ r\sqrt{3}}{3} = \frac{4 \times 127\ \text{pm} \times \sqrt{3}}{3} = 293\ \text{pm}$$

Since the edge of one unit cell is 293 pm or 2.93 × 10⁻⁸ cm, the volume of one unit cell is (2.93 × 10⁻⁸ cm)³ or 2.52 × 10⁻²³ cm³.

Each unit cell in a body-centered cubic arrangement contains an equivalent of 2 atoms so that there are 3.01 × 10²³ unit cells in one molar mass of chromium.

The volume occupied by one molar mass of Cr = 2.52 × 10⁻²³ cm³/unit cell × 3.01 × 10²³ unit cells = 7.59 cm³.

The density of Cr = $\dfrac{M}{V}$ = $\dfrac{52.0 \text{ g}}{7.59 \text{ cm}^3}$ = 6.85 g/cm³

Example No. 13-11. CsI crystallizes in body-centered cubic arrangement with the cesium ions at each corner of the cube and an iodide ion in the center of the cube. Given the edge of the CsI unit cell as 445 pm, calculate:

(a) the Cs-Cs distance

(b) the I-I distance

(c) The Cs-I distance

(d) the ionic radius of iodine given the ionic radius of cesium as 169 pm.

    (a) Since the unit cell contains a cesium ion at each corner, and since the edge of a unit cell is 445 pm, the distance between cesium ions is 445 pm.

    (b) Since the iodide ions are located at the center of the cube, the distance from the iodide ion in one cube to the iodide ion in the adjacent cube is equal to the length of a unit cell or 445 pm.

    (c) To calculate the Cs-I distance

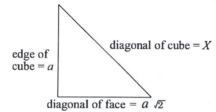

$$X^2 = a^2 + (a\sqrt{2})^2 = 3a^2$$
$$X = a\sqrt{3} = 445 \text{ pm} \times \sqrt{3} = 770 \text{ pm}$$

The diagonal of a body-centered cube of CsI has a length equal to the radius of one cesium ion, twice the radius of the iodide ion, and again the radius of a cesium ion, or

$$X = 2(\text{radius}_{Cs^+} + \text{radius}_{I^-})$$

$$770 \text{ pm} = 2 (\text{radius}_{Cs^+} + \text{radius}_{I^-})$$

$$385 \text{ pm} = (\text{radius}_{Cs^+} + \text{radius}_{I^-})$$

Thus, the cesium and iodide distance is 385 pm.

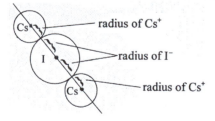

    (d) Given that the radius of Cs^+ is 169 pm, the radius of the I^- ion is 385 pm $-$ 169 pm $=$ 216 pm.

Example No. 13-12. What is the simplest formula for a compound whose atoms form a body-centered cubic arrangement in which atoms of X occupy the corners of the unit cell and atoms of Y the centers of the body?

    Each corner atom of body-centered cubic unit cell contributes one-eighth of itself to its unit cell. The total contribution of all eight corner atoms is one so that the unit cell contains the equivalent of one atom of X.

    The atom in the center of the body-centered cube belongs wholly to that unit cell so that the unit cell contains one atom of Y.

    Therefore, the simplest formula of the compound is XY.

III.   THE FACE-CENTERED CUBE

Consider a cubic unit cell which has an atom (or ion) at each corner and also an atom (or ion) at the center of each face.

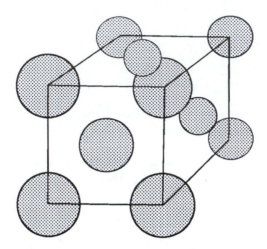

Such a cube is called a face-centered cube.

Example No. 13-13. How many atoms are present in each unit cell of face-centered cube?

As in the simple cube, each corner atom contributes one-eighth of itself to that cube. All eight corner atoms contribute a total of one atom to that unit cell.

An atom in the center of a face belongs equally to two adjacent unit cells. That is, the atom in the center of a face contributes only 1/2 of itself to each unit cell. Since there are six faces to a cube, the total contribution of all of the faces of the face-centered cube equals 3 atoms.

Thus, there is an equivalent of 4 atoms present in each unit cell of face-centered cube.

Example No. 13-14.   Assuming that the corner atoms of a face-centered cube are in direct contact with the atom in the center of a face, and also assuming that the atoms are all of the same size, what percent of the face-centered cube is empty space?

Consider the following diagram of one face of face-centered cube of edge a.

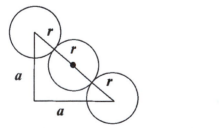

$$a^2 + a^2 = (4\,r)^2$$
$$2\,a^2 = 16\,r^2$$
$$a^2 = 8\,r^2$$
$$a = 2\,r\sqrt{2}$$

The volume of one unit cell of the cube = $(a)^3 = (2\,r\sqrt{2})^3$
$$= 16\,r^3\sqrt{2}$$

The volume of the 4 (equivalent) atoms present in one unit cell of a face-center cube =
$4\,(4/3\,\pi\,r^3) = 16/3\,\pi\,r^3$.

The amount of empty space = $16\,r^3\,\sqrt{2} - 16/3\,\pi\,r^3$ and the percent of empty space =

$$\frac{16\,r^3\,\sqrt{2} - 16/3\,\pi\,r^3}{16\,r^3\,\sqrt{2}} \times 100 = 26.0\%$$

Example No. 13-15. The atomic radius of calcium is 197 pm. What is the density of calcium if it forms a face-centered cube?

The diagonal of a face of a face-centered cube has a length equal to *4 r* so

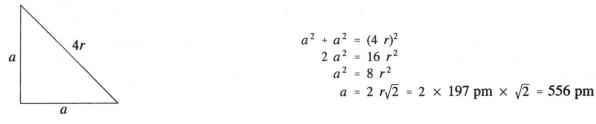

$$a^2 + a^2 = (4\ r)^2$$
$$2\ a^2 = 16\ r^2$$
$$a^2 = 8\ r^2$$
$$a = 2\ r\sqrt{2} = 2 \times 197\ \text{pm} \times \sqrt{2} = 556\ \text{pm}$$

The volume of a unit cell of calcium = a^3 = (556 pm)^3 = (5.56 × 10^{-8} cm)^3 = 1.72 × 10^{-22} cm^3.

A unit cell of a face-centered cube contains an equivalent of four atoms so that 4 Ca atoms occupy a volume of 1.72 × 10^{-22} cm^3.

1 molar mass of Ca, 40.1 g, contains 6.02 × 10^{23} atoms,

so 4 atoms of Ca have a mass of $\dfrac{40.1\text{g}}{6.02 \times 10^{23}}$ × 4 = 2.66 × 10^{-22} g.

The density of Calcium (the density of a unit cell) $= \dfrac{M}{V} = \dfrac{2.66 \times 10^{-22}\ \text{g}}{1.72 \times 10^{-22}\ \text{cm}^3} = 1.55\ \text{g/cm}^3$

Example No. 13-16. Aluminum forms a face-centered cube with an edge of a unit cell equal to 404 pm. Calculate (a) the atomic volume of aluminum, (b) the density of aluminum, (c) the atomic radius of aluminum.

(a) The volume of one unit cell of aluminum = a^3 = (404 pm)^3 = (4.04 × 10^{-8} cm)^3 = 6.59 × 10^{-23} cm^3.

One molar mass of aluminum contains 6.02 × 10^{23} atoms, and also $\dfrac{6.02 \times 10^{23}}{4}$ or 1.505 10^{23} unit cells (4 aluminum atoms per unit cell in a face-centered cube).

The volume occupied by one molar mass of aluminum (its atomic volume) = 6.59 × 10^{-23} cm^3/unit cell × 1.505 10^{23} unit cells = 9.92 cm^3.

(b) The density of aluminum $= \dfrac{M}{V} = \dfrac{27.0\ \text{g}}{9.92\ \text{cm}^3} = 2.72\ \text{g/cm}^3$.

(c) Assuming that the aluminum atoms are in contact along the diagonal of a face,

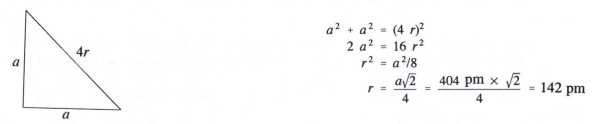

$$a^2 + a^2 = (4\ r)^2$$
$$2\ a^2 = 16\ r^2$$
$$r^2 = a^2/8$$
$$r = \dfrac{a\sqrt{2}}{4} = \dfrac{404\ \text{pm} \times \sqrt{2}}{4} = 142\ \text{pm}$$

Example No. 13-17. What is the simplest formula for a compound containing *X* ions at the corners of a face-centered cube and *Y* ions at the center of each face?

Each corner atom or ion in a cubic arrangement contributes one-eighth of itself to the unit cell. That is, the total contribution of all of the corner ions is equivalent to one ion. So the unit cell contains one *X* ion.

Each Y ion, in the center of a face, contributes one-half of itself to the unit cell. From the 6 faces the total contribution of all of the Y ions is equivalent to three Y ions.

Therefore, the formula of the compound is XY_3

Arrangements of Sodium and Chloride Ions in NaCl

We have mentioned previously that NaCl is considered to have a simple cubic arrangement of Na^+ and Cl^- ions indicated below:

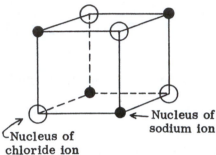

However, NaCl is frequently considered as having the sodium ions themselves in the shape of a face-centered cube and the chloride ions themselves as having a face-centered cubic arrangement, as shown in the following diagram:

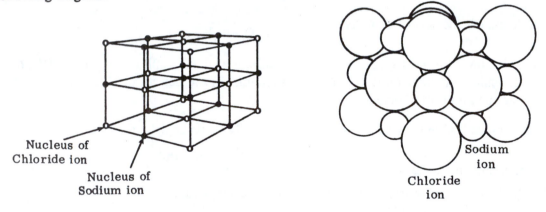

Example No. 13-18. Calculate the number of sodium ions and the number of chloride ions present in a unit cell of the above face-centered cubic arrangement of ions. What is the length of an edge in such a unit cell?

Consider the chloride ions at the corners. Each one contributes one-eighth of itself to that unit cell. The total contribution of all of the corner ions is one chloride ion. Each chloride ion in the center of a face contributes one-half of itself to that cell. The total contribution of all of the chloride ions in the (six) faces is equivalent to three chloride ions. Therefore, the total equivalent number of chloride ions present in this unit cell is 4.

Each sodium ion along the edge of the cube contributes only one-fourth of itself to that cell. Since there are 12 sodium ions present along the edges of this cell, the equivalent number of sodium ions present because of these ions along the edge is 3. In addition, there is also one sodium ion present at the center of the cube. This ion belongs wholly to that cell. That is, the total number of equivalent sodium ions present in this unit is 4.

Thus in the unit cell of the above type, there are 4 sodium ions and 4 chloride ions or an equivalent of 4 NaCl units.

The length of an edge of such a cube, as shown in the above diagram, is equal to twice the Na^+Cl^- distance.

Problem Assignment

13-1.   Manganese crystallizes in a simple cubic shape with an atomic radius of 116 pm. What is the theoretical density of Mn? . . . . (7.30 g/cm^3)

13-2.   Given the observed density of Mn as 7.19 g/cm^3, calculate the percent of empty sites in the manganese crystal.

13-3.   Sodium bromide crystallizes in a simple cubic shape with a density of 3.20 g/cm^3. Calculate the length of a unit cell. . . . . (299 pm)

13-4.   Given the ionic radius of a sodium ion as 95 pm, calculate the ionic radius of a bromide ion using the data in problem 13-3.

13-5.   A unit cell of a simple cubic structure consists alternately of X and Y ions. What is the simplest formula of the compound? . . . . (XY)

13-6.   Phosphorus crystallizes with a simple cubic shape whose "a" is 300 pm. Calculate the density of phosphorus and also the percent vacant sites if the observed density is 1.82 g/cm^3.

13-7.   LiCl crystallizes in a simple cubic shape. Given the density of LiCl as 2.07 g/cm^3 and the molecular mass as 42.4, calculate the molecular volume of LiCl. If the Li^+ ion is 1/3 the size of the Cl^- ion , calculate the size of each. . . . . (20.5 cm^3
64 pm for Li^+,
193 pm for Cl^-)

13-8.   Titanium monoxide crystallizes in a simple cubic shape with an observed density of 4.93 g/cm^3. If TiO contains 12.0% vacant sites, what is the theoretical density of TiO? What is the length of a unit cell?

13-9.   The element potassium crystallizes as a body-centered cube. If the atomic radius of K is 230 pm, what is the length of a unit cell? What is the density of K?   . . . . (531 pm, 0.867 g/cm^3)

13-10.   The density of vanadium is 6.10 g/cm^3. Calculate the atomic radius of vanadium, assuming that it crystallizes as a body-centered cube.

13-11.   The element tantalum crystallizes as a body-centered cube. If the length of a unit cell is 331 pm, and if the density is 16.6 g/cm^3, calculate the value of Avogadro's number.
. . . . (6.01 $\times 10^{23}$)

13-12.   An element has a density of 1.53 g/cm^3. It crystallizes as a body-centered cube with the length of a unit cell equal to 570 pm. Calculate the molecular mass of the element. Assuming a light error due to vacant sites, identify the element.

13-13.   Europium crystallizes as a body-centered cube. If its density is 5.26 g/cm^3, calculate the atomic radius of Eu. . . . . (198 pm)

13-14.   Platinum crystallizes as a body-centered cube. If the atomic radius of Pt is 134 pm, calculate the length of a unit cell and the density of Pt.

13-15.   An element has a density of 19.3 g/cm^3. It crystallizes in a body-centered cubic structure. The length of a unit cell is 317 pm. Is the element W, Au, or Np? . . . . (W)

13-16.   If the observed density of Pt is 21.4 g/cm^3, calculate the percent of vacant sites from the data in problem 13-14.

13-17. Strontium crystallizes in a face-centered cubic arrangement. If the atomic radius of strontium is 210 pm, calculate the density of this element. . . . . (2.78 g/cm^3)

13-18. Given the density of strontium as 2.60 g/cm^3, calculate the percent of empty sites using the data in problem 13-17.

13-19. Silver crystallizes in a face-centered cubic system with the length of a unit cell equal to 409 pm. Calculate the density of silver and its atomic radius. . . . . (10.5 g/cm^3, 145 pm)

13-20. A certain element has a density of 6.67 g/cm^3. It crystallizes in a face-centered cubic system with the length of a unit cell equal to a 519 pm. Calculate the molar mass of the element. What is its identity?

13-21. Calcium oxide forms a sodium chloride type structure. If the effective radius of the calcium ion is 110 pm and the effective radius of the oxide ion is 151 pm, calculate the density of CaO. . . . . (2.62 g/cm^3)

13-22. Lead forms a face-centered cubic arrangement with a density of 11.4 g/cm^3. If the atomic radius of lead is 170 pm, calculate the percent of empty sites.

13-23. An element forms a face-centered cubic system with a density of 12.4 g/cm^3. If the atomic radius is 134 pm, calculate the molar mass and the identity of the element. . . . . (103 g, Rh)

13-24. Barium oxide crystallizes in a sodium chloride face-centered structure. Its density is 5.72 g/cm^3. Calculate the Ba-O distance. If the barium and the oxide ions have approximately the same size, calculate the ionic radius of each.

Chapter 14
THE PERIODIC CHART

The periodic chart contains a great deal of information for the chemist. We shall discuss those facts which are important to the beginner and leave the others to the more advanced student. There is a square on the chart for each element, and this square contains information about the element.

I. ATOMIC NUMBER, ATOMIC MASS, SYMBOL

The atomic number is located in the upper corner of the square—right or left as the case may be. The atomic mass is directly below the symbol of the element which is the large letter (or letters) in the center of the square.

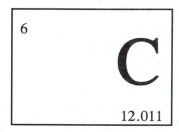

represents Carbon, symbol C,

atomic number 6, atomic mass 12.011.

II. PHYSICAL STATE

On some charts, the color of the symbol designates the state of the element at room temperature—a black symbol means a solid; a blue symbol, a liquid; and an orange symbol, a gas.

III. GROUPS AND PERIODS

Some periodic charts use a slightly different numbering system for groups than do other charts. Note that in the periodic chart inside the back cover that the groups are labelled IA, IIA, IIIB to VIIIA and also 1-18. Selection of either numbering system is a matter of choice and either system may be found in textbooks or on wall charts. In this book we will refer to both systems such as: VIIB (17). Group IA (1) contains the elements H, Li, Na, K, etc., and Group VII A (17) contains F, Cl, Br, I, etc.

The number along the side of the chart designates the period. The elements in period 2 are Li, Be, B, C, N, O, F, Ne and those in period 3 begin with Na, Mg, . . . and end with Ar.

IV. GROUP AND OXIDATION NUMBER

The arrangement of the electrons of the elements (metals) in Group IA (1) is as follows:

H)	1				
Li)	2)	1			
Na)	2)	8)	1		
K)	2)	8)	8)	1	
Rb)	2)	8)	18)	8)	1
Cs)	2)	8)	18)	18)	8) 1

We note that all of these elements have one electron in their outer energy level, the valence shell. This electron can be lost very easily and when lost gives these elements a charge of +1. The oxidation number is the same as the charge on the ion, +1.

In Group IIA (2), we see the metals

Be) 2) 2   Ca) 2) 8) 8) 2

Mg) 2) 8) 2   Sr) 2) 8) 18) 8) 2

Here all the elements have a charge of +2 when they lose their two outer (valence) electrons in the formation of a compound. Their oxidation number is +2.

In Group VIIA (17) are the non-metals.

F) 2) 7,   Cl) 2) 8) 7, etc.

These elements will acquire 1 electron to complete their outer energy level to 8, thus they have an ionic charge of -1 and likewise an oxidation number of -1.

Therefore, the elements in Group IA (1) have an oxidation number of +1; in Group IIA (2), an oxidation number of +2; and in Group VIIA (17), an oxidation number of -1. If we consider all the groups in the same manner, we will find that:

a)   For metals, THE A-GROUP NUMBER IS THE SAME AS THE OXIDATION NUMBER.
b)   For non-metals, THE A-GROUP NUMBER MINUS EIGHT GIVES THE OXIDATION NUMBER.

In some groups both types of oxidation numbers are possible.

GROUP #	IA(1)	IIA (II)	IIIA (13)	IVA (14)	VA (15)	VIA (16)	VIIA (17)	VIIIA(18)
OXIDATION NUMBER	+1	+2	+3	+4,-4	+5,-3	+6,-2	+7,-1	0

From this table, we can find the formula of the oxygen compound of any metal and the formula for the hydrogen compound of any non-metal. Thus, the oxygen compound of Li (Group IA (1), oxidation number +1) is Li_2O; for Ba (Group IIA (2), oxidation number +2) BaO; and for Sn (Group IVA (14), oxidation number +4) SnO_2.

The hydrogen compound of Carbon, (Group IVA (14), oxidation number -4) is CH_4; for P (Group VA(15), oxidation number -3) PH_3; and for S (Group VIA (16), oxidation number -2) SH_2 or H_2S.

V. METALS AND NON-METALS

The metals are located on the left side of the heavy line on the chart while the non-metals are located on the right.

VI. COLOR IN A GROUP

As we go down a group, the colors of the elements generally become darker. In Group VIIA (17),

> F is a pale greenish-yellow gas.
> Cl is a yellow-green darker gas.
> Br is a brown-red liquid.
> I is a purple-black solid.

In general, then, those elements at the bottom of a group are darker in color than those at the top.

VII. DENSITY IN A GROUP

As we go down a group, the atomic mass increases much more rapidly than the atomic volume so that the density increases.

VIII. ACTIVITY AND IONIZATION ENERGY IN A GROUP

As we go down Group IA (1), the one outer electron is much farther from the nucleus and so is more easily removed. This means that the elements become more active as metals due to the greater ease in losing electrons, and it also means that the ionization energy decreases for the same reason.

In Group VIIA (17), the outer 7 electrons are progressively farther from the nucleus and so the ability to capture an electron decreases. Therefore, the activity as a non-metal decreases as we go down Group VIIA (17). The elements at the top of Group VIIA (17) will hold to their outer electrons more tightly than those at the bottom and so the ionization energy will decrease as we go down this group. Thus we can generalize:

on the left side (the metals)—the ionization energy decreases down a group.

on the right side (non-metals)—the ionization energy decreases down a group.

on the left side (metals)—the activity increases down a group.

on the right side (non-metals)—the activity decreases down a group.

We can state this in another way—the elements in Group VIIA (17) are more electronegative than the elements in Group IA (1). The most electronegative element is at the top of Group VII (17) and the least electronegative (most electropositive) is at the bottom of Group IA (1).

IX. ACTIVITY ACROSS A PERIOD

As we go across a period, the activity of the metals decreases and the activity of the non-metals increases. Thus, in the period 3, Na is a stronger metal than Mg, etc., and Cl is a stronger non-metal than S.

X. ATOMIC RADIUS

As we go down a group, the atomic radius increases because more energy levels are being added and these energy levels are farther from the nucleus.

As we go across a period, we are adding electrons to the same energy level. The positive charge on the nucleus is increasing at the same time, and this increased positive charge pulls the electron energy levels in closer so that the atomic radius decreases across a period.

XI. IONIC RADIUS

When a metal loses an electron, the positive charge on the nucleus is greater than the negative charge in the energy levels so the nucleus pulls in the energy levels and decreases the radius. Thus, for metals, the ionic radius is less than the atomic radius.

When non-metals gain electrons, the positive charge on the nucleus cannot hold these electrons so tightly and so the ionic radius for non-metals is greater than the atomic radius.

XII. TRANSITION ELEMENTS

If we look at the periodic chart, we see that there is a group of elements between Group IIA and Group IIIA. These elements are numbered IIIB to IIB (3-12) and are called transition elements. Transition elements than one type of ion because they can lose electrons from an inner energy level as well as from the outer energy level.

XIII. THE PERIODIC LAW

The periodic law states that "the properties of the elements are a periodic function of their atomic number." This means that certain elements exhibit similar properties, and that these similarities occur periodically. Thus, Li (atomic number 3) is very similar to Na (atomic number 11) which is similar to K (atomic number 19) and so on.

We note that the atomic numbers increase directly in order from 1 to 118 on the chart. The atomic masses generally increase as we go through the chart but there are some exceptions to this rule (due to the presence of isotopes). They are:

Argon	atomic number 18	atomic mass 39.948
Potassium	atomic number 19	atomic mass 30.0983
Cobalt	atomic number 27	atomic mass 58.9330
Nickel	atomic number 28	atomic mass 58.69
Tellurium	atomic number 52	atomic mass 127.60
Iodine	atomic number 53	atomic mass 126.9043

Atomic and Ionic Sizes of Some Elements

Angstrom units
1 2 3 4 5

Problem Assignment

Which is more active?

14-1.   Ba or Ca
14-2.   F or Cl
14-3.   Mg or Ca
14-4.   O or S
14-5.   Br or I
14-6.   Ne or Kr
14-7.   Which are metals? K, Al, S, Ne, Ca, C, Mn, Sc, W, Se, P, U.

14-8. How does the ionization energy vary
   a) in Group IIA (2)?
   b) in Period 3?
   c) from the elements in Group IA (1) to the corresponding elements in Group VIIA (17)?

14-9. How does the activity vary
   a) in Group IA (1)?
   b) in Group VIA (16)?
   c) in Period 2?

14-10. How does the atomic radius vary in a period? in a group?

14-11. How does the ionic radius vary in a period? in a group?

14-12. What is the most electronegative element?

14-13. What is the most electropositive element?

14-14. What is the formula for an oxide of each of the following elements?
   a) Carbon, atomic number 6.
   b) Magnesium, atomic number 12.
   c) Sodium, atomic number 11.
   d) Barium, atomic number 56.
   e) Lead, atomic number 82.
   f) Aluminum, atomic number 13.
   g) Radium, atomic number 88.
   h) Silicon, atomic number 14.
   i) Cesium, atomic number 55.
   j) Phosphorus, atomic number 15.

14-15. What is the formula of the Hydrogen compound of each of the following elements?
   a) Bromine, atomic number 35.
   b) Phosphorus, atomic number 15.
   c) Carbon, atomic number 6.
   d) Antimony, atomic number 51.
   e) Silicon, atomic number 14.
   f) Selenium, atomic number 34.
   g) Nitrogen, atomic number 7.
   h) Astatine, atomic number 55.
   i) Iodine, atomic number 53.
   j) Oxygen, atomic number 8.

14-16. What is the formula of the compound formed from the reaction between:
   a) Sulfur and Potassium.
   b) Aluminum and Fluorine.
   c) Magnesium and Phosphorus.
   d) Radium and Sulfur.
   e) Barium and Bromine.
   f) Indium and Fluorine.
   g) Calcium and Sulfur.
   h) Potassium and Phosphorus.
   i) Aluminum and Sulfur.
   j) Titanium and Chlorine.

Chapter 15

THE VSEPR THEORY

The VSEPR (valence shell electron-pair repulsion) theory is based upon the idea that electron pairs in the valence shell of an atom repel one another. A molecule (or ion) will have the lowest potential energy when its electron pairs are in such a geometric position as to minimize their repulsions. The electrons try to get as far from each other as possible. The relative locations of the electrons and of the atoms accounts for the molecular geometry of the molecule. The VSEPR theory is useful in predicting approximate shape of molecules (or ions) formed from non-metals.

The following method may be used in determining the shapes of molecules (or polyatomic ions):

1. Draw the electron dot structure and stick structure of the compound.

2. Count the number of atoms bonded to the central atom and also the number of nonbonded electron pairs on the central atom.

3. Use Table 15-1 to predict the shape of the molecule (or ion).

Table 15-1—BONDING AND SHAPE OF MOLECULES AND IONS

Number of atoms bonded to central atom	Number of pairs of nonbonded electrons on central atom	Total	Shape of molecule or ion
2	0	2	linear
2	1	3	bent
2	2	4	bent
3	1	4	pyramidal
4	0	4	tetrahedral
4	2	6	square planar
5	0	5	trigonal bipyramidal
6	0	6	octahedral

Example No. 15-1. Determine the shape of CH_4

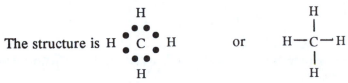

The structure is

Number of atoms bonded to central atom = 4; number of nonbonded electron pairs on central atom = 0; total 4 + 0 = 4.

Therefore, predicted shape is tetrahedral.

Example No. 15-2.   Determine the shape of the SF$_6$ molecule.

The structure is 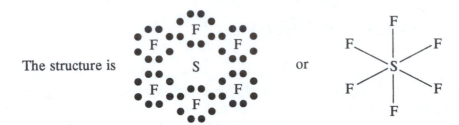   or

Since there are 6 atoms bonded to the central atom, and no nonbonded electron pairs on the central atom, the total is 6 and the shape predicted is octahedral.

Example No. 15-3.   What is the shape of the NH$_3$ molecule?

The structure is   H $\cdot\cdot$ N $\cdot\cdot$ H   or   H—N—H

                          H   H

There are 3 atoms bonded to the central atom and 1 pair of nonbonded electrons on the central atom. 3 + 1 = 4 and the shape predicted will be pyramidal.

Example No. 15-4.   Predict the shape of H$_2$O.

The structure is   H $\cdot\cdot$ O $\cdot\cdot$ H   or   H—O—H   or   H—O $\cdot\cdot$

                                                                                   H

The number of atoms bonded to the central atom = 2; the number of nonbonded electron pairs on the central atom = 2; total = 4

Therefore, shape should be bent.

Example No. 15-5.   What is the shape of ClO$_3^-$.

The structure is    or   O—Cl—O]$^-$

                                                              O

Total = 3 (three atoms bonded to central atom) + 1 (one electron pair) = 4. Therefore, the structure will be pyramidal.

Example No. 15-6.   Determine the shape of the CO$_2$ molecule.

The structure is   O $\cdot\cdot$ C $\cdot\cdot$ O   or   O=C=O

Number of atoms bonded to the central atom = 2; number of nonbonded electron pairs on central atom = 0; total = 2 so shape is predicted to be linear. Note that the presence of double (or triple) bonds does not affect the number of atoms bonded to the central atom.

Example No. 15-7. Determine the shape of the NO_2^- ion.

The structure is 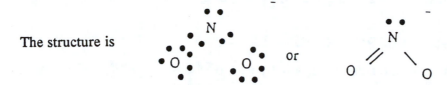 or

The number of atoms bonded to the central atom is 2; the number of unbonded electron pairs on the central atom is 1. Thus $2 + 1 = 3$ so that the predicted shape is bent.

Problem Assignment

Predict the shape of the following compounds or ions.

15-1. BCl_3 .... (3 + 0 = 3; trigonal)

15-2. PBr_5 .... (5 + 0 = 5; trigonal bipyramidal)

15-3. ClF_3 .... (4 + 0 = 4; tetrahedral)

15-4. SCl_2 .... (2 + 2 = 4; bent)

15-5. $BeCl_2$

15-6. $AsBr_5$

15-7. SO_3^{2-}

15-8. PO_3^{2-}

15-9. CI_4

15-10. AsF_6^-

Chapter 16

OXIDATION REDUCTION EQUATIONS

I. BALANCING BY CHANGE IN OXIDATION NUMBER

Before we can begin to balance oxidation-reduction (redox) reactions we must define the terms themselves.

An atom which has been *oxidized* or has undergone oxidation has lost electrons. It has had an algebraic increase in oxidation number.

An atom which has been *reduced* or undergone reduction has gained electrons. It has had an algebraic decrease in oxidation number.

To balance a redox equation use the following rules:

1. Write the complete equation, listing the oxidation numbers of all the elements, and locate those elements which are changing their oxidation number.

2. Calculate the change in oxidation number and indicate this change below the equation.

3. Multiply the change in oxidation number for each atom by the total number of those atoms which are changing in oxidation number.

4. Find a least common multiple for the total increase and decrease in oxidation number.

5. Equate the increase and decrease in oxidation numbers since they must always be equal.

6. Complete the remainder of the equation by adding the proper coefficients to make sure that there are the same number of each type of atom on each side of the equation.

7. Note that any element in the free state, that is, uncombined, has an oxidation number of zero.

Example No. 16-1. Balance: $H_2 + O_2 \rightarrow H_2O$

step 1. Write in all the oxidation numbers.

$$\overset{0}{H_2} + \overset{0}{O_2} \rightarrow \overset{2(+1)}{H_2} \overset{-2}{O}$$

(Since the molecules of hydrogen and oxygen are uncombined, their atoms have an oxidation number of zero; from the previous chapter on oxidation numbers we know that the oxidation number of oxygen in water is -2 and that of each hydrogen is $+1$.)

step 2. Show the increase and decrease in oxidation numbers.

Since the hydrogen is changing from an oxidation number of 0 to $+1$, there is an increase of 1 in oxidation number. The oxygen, in changing from an oxidation number of 0 to -2, decreases by 2 in oxidation number. These changes may be indicated in the equation as:

$$\overset{0}{H_2} + \overset{0}{O_2} \rightarrow \overset{2(+1)}{H_2} \overset{-2}{O}$$

↑ 1 for each atom ↓ 2 for each atom

where the upward pointing arrow indicates an increase in oxidation and the downward pointing arrow a decrease in oxidation number.

step 3. Find the total change in oxidation number for each of the elements that is changing.

The total increase in oxidation number for the hydrogen is 2 (1 for each of the hydrogens). The total decrease in oxidation number for the oxygen is 4 (2 for each of the oxygens). These total changes in oxidation number may be indicated as:

Total changes:

$$\overset{0}{H_2} + \overset{0}{O_2} \rightarrow \overset{2(+1)\ -2}{H_2\ \ \ O}$$

$$\uparrow 2 \qquad\qquad \downarrow 4$$

step 4. The least common multiple for an increase of 2 and a decrease of 4 is 4.

step 5. Equate the increase in oxidation number with the decrease.

The increase in oxidation number for the hydrogen (2) goes into the LCM (least common multiple) twice so we place a 2 in front of the H_2. The decrease in oxidation number for the O_2 (4) goes into the LCM once so we place a 1 in front of the O_2 (no number in front of a formula designates 1). So we have:

$$2H_2 + O_2 \rightarrow H_2O$$

where the total increase in oxidation number is now equal to the total decrease.

step 6. Complete the equation.

Since we now have a total of 4 H's on the left side of the equation, we must have 4 H's on the right side so we place a 2 in front of the H_2O and then we have 4 H's on each side of the equation, or

$$2H_2 + O_2 \rightarrow 2H_2O$$

We note that there are also 2 O's on each side of the equation so that it is completely balanced.

After you have tried several redox equations you will be able to combine many of these steps and save time in balancing.

Let us see what was OXIDIZED and what was REDUCED. Since the definition of oxidation is an increase in oxidation number, the H_2 must have been oxidized because its oxidation number did increase (from 0 to 1 each). Also, the O_2 must have been reduced since it decreased in oxidation number (from 0 to −2 each).

There are two more terms we should know—oxidizing agent and reducing agent. AN OXIDIZING AGENT OXIDIZES SOMETHING AND IT ITSELF IS REDUCED; A REDUCING AGENT REDUCES SOMETHING AND IT ITSELF IS OXIDIZED. Thus the H_2 was oxidized. What oxidized it? The O_2; so the O_2 is the oxidizing agent. The O_2 was reduced. What reduced it? The H_2; so the H_2 is the reducing agent.

$$\begin{array}{cc} H_2 & O_2 \\ \text{oxidized} & \text{reduced} \\ \text{reducing agent} & \text{oxidizing agent} \end{array}$$

Example No. 16-2. Balance: $KClO_3 \rightarrow KCl + O_2$

Listing the oxidation number according to the methods discussed in the previous chapter we have:

$$\overset{+1}{K}\ \overset{+5}{Cl}\ \overset{3(-2)}{O_3} \rightarrow \overset{+1}{K}\ \overset{-1}{Cl} + \overset{0}{O_2}$$

We note that the Cl and the O are changing in oxidation number. The Cl, in changing from +5 to −1 exhibits a decrease of 6 in oxidation number. The O, in changing from −2 to 0 exhibits an increase of 2 in oxidation number for each of the O atoms present. Since there are three O atoms present, the total increase in oxidation number is 6. These changes may be indicated as:

Total changes:
$$\overset{\underset{\downarrow 6}{+5}}{K \quad Cl} \quad \overset{3(-2)}{O_3} \quad \underset{\uparrow 6}{\longrightarrow} \quad K \quad \overset{-1}{Cl} \quad + \quad \overset{0}{O_2}$$

Since the increase in oxidation number here does equal the decrease, we need no LCM and so leave the left side of the equation as is.

Now to complete the equation. Since we have 1 $KClO_3$ on the left side of the equation we can have only 1 KCl on the right because the number of K's on each side of the equation must be the same and so must the number of Cl's. Since we have 3 O's on the left side of the equation, we must have 3 O's on the right side and we get these by placing a 1½ in front of the O_2 (1½ × O_2 = 3 O's). Thus we have

$$KClO_3 \rightarrow KCl + 1½ \, O_2$$

However, we can never have fractions of atoms or molecules in a chemical equation so we multiply the whole equation by 2 to get

$$2 \, KClO_3 \rightarrow 2 \, KCl + 3 \, O_2$$

which is balanced since it contains the same number of each type of atom on each side of the equation.

WHAT WAS REDUCED? An error made by many students is the statement that the Chlorine was reduced since it decreased in oxidation number. It is true that a decrease in oxidation number is reduction, but the chlorine did not decrease in oxidation number, the Cl^{+5} decreased. This makes quite a difference as we shall see in the following paragraph.

WHAT WAS OXIDIZED? The O^{-2} was oxidized since it increased in oxidation number. If we had said that the Oxygen was oxidized that would have been incorrect because Oxygen is a product of the reaction and not one of the original reactants.

Also the Cl^{+5}, since it was reduced, is the oxidizing agent and the O^{-2} which was oxidized is the reducing agent.

Example No. 16-3. Balance: $P + HNO_3 + H_2O \rightarrow NO + H_3PO_4$

We will combine a few of the steps to save space. Therefore, we have

Total changes:
$$\underset{\uparrow 5}{\overset{0}{P}} + \overset{+1}{H} \underset{\downarrow 3}{\overset{+5}{N}} \overset{3(-2)}{O_3} + \overset{2(1)}{H_2} \overset{-2}{O} \rightarrow \overset{+2}{N} \overset{-2}{O} + \overset{3(+1)}{H_3} \overset{+5}{P} \overset{4(-2)}{O_4}$$

The LCM between an increase of 5 and a decrease of 3 is 15. The increase in oxidation number by the P (5) goes into the LCM 3 times so we place a 3 in front of the P; the decrease in oxidation number by the N (3) goes into the LCM 5 times so we place a 5 in front of the HNO_3 (to give us 5 N's) or

$$3 \, P + 5 \, HNO_3 + (\;) \, H_2O \rightarrow (\;) \, NO + (\;) \, H_3PO_4$$

3 P's on the left side of the equation must give 3 P's on the right so we put a 3 in front of the H_3PO_4 to give us a total of 3 P's there. 5 HNO_3 on the left side of the equation contain 5 N's so we must have 5 N's on the right side and we get these by placing a 5 in front of the NO; or we have

$$3 \, P + 5 \, HNO_3 + (\;) \, H_2O \rightarrow 5 \, NO + 3 \, H_3PO_4$$

Now to complete the equation, let us balance the H's. There are 9 H's on the right side of the equation (3 H_3). On the left side there are 5 H's in the HNO_3's so we need 4 more and we get these by taking 2 H_2O (4 H's), or

$$3 P + 5 HNO_3 + 2 H_2O \rightarrow 5 NO + 3 H_3PO_4$$

To check, let us see if the O's balance. There are $5 \times 3 + 2 \times 1 = 17$ O's on the left side and $5 \times 1 + 3 \times 4 = 17$ O's on the right side, so the equation is correctly balanced.

WHAT WAS OXIDIZED? The P since it increased in oxidation number. Thus it is the reducing agent.

WHAT WAS REDUCED? The N^{+5} (or we could say the HNO_3) was reduced since it decreased in oxidation number and, therefore, it is the oxidizing agent.

Example No. 16-4. Balance: $MnO_2 + HCl \rightarrow MnCl_2 + Cl_2 + H_2O$

First let us put in the oxidation number and see what is changing.

$$\begin{array}{ccccccc} +4 \ 2(-2) & +1 \ -1 & +2 \ 2(-1) & 0 & 2(+1) \ -2 \\ Mn \ O_2 & + \ H \ Cl \rightarrow & Mn \ Cl_2 & + \ Cl_2 & + \ H_2 & O \end{array}$$

We note that the Mn is changing from +4 to +2 and so is decreasing by 2 in oxidation number. Some of the Cl^{-1} is going from –1 to 0 and so is increasing by 1 in oxidation number. We must remember that not all the Chlorine is changing in this equation. So we have

$$\begin{array}{ccccc} +4 & -1 & +2 & 0 \\ Mn \ O_2 & + \ H \ Cl \rightarrow & Mn \ Cl_2 & + \ Cl_2 & + H_2O \\ \downarrow 2 & \uparrow 1 \end{array}$$

The LCM for an increase of 1 and a decrease of 2 is 2. The total increase in oxidation number (1) goes into the LCM twice so we place a 2 in front of the HCl. The total decrease in oxidation number by the Mn of the MnO_2 (2) goes into the LCM once so we place a 1 in front of the MnO_2 (the number 1 being understood if no coefficient is given) or

$$MnO_2 + 2 HCl \rightarrow (\ ) MnCl_2 + (\ ) Cl_2 + (\ ) H_2O$$

1 MnO_2 can give only 1 $MnCl_2$ since 1 Mn on the left side of equation can give only 1 Mn on the right side. 2 HCl's can give 1 Cl_2 since both of these Cl's are changing in valence. So we write

$$MnO_2 + 2 HCl \rightarrow MnCl_2 + Cl_2 + (\ ) H_2O$$

We notice that 2 Cl's in the $MnCl_2$ did not change in oxidation number so we must add 2 more HCl's to the left side since the original 2 HCl's consist only of the chlorines which did change in oxidation number. Therefore, we need 2 + 2 or 4 HCl on the left side of the equation, or

$$MnO_2 + 4 HCl \rightarrow MnCl_2 + Cl_2 + (\ ) H_2O$$

To complete the equation we see that we have 4 H's on the left side of the equation so we place a 2 in front of the H_2O to give 4 H's on the right. There are 2 O's on each side of the equation so that the properly balanced result is

$$MnO_2 + 4 HCl \rightarrow MnCl_2 + Cl_2 + 2 H_2O$$

WHAT WAS OXIDIZED? The Cl^- (not the chlorine) was oxidized since it increased in oxidation number (or the HCl was oxidized). Therefore, the Cl^- or the HCl is the reducing agent.

WHAT WAS REDUCED? The Mn^{+4} or the MnO_2 decreased in oxidation number and so was reduced and, therefore, is the oxidizing agent.

Example No. 16-5. Balance:

$$FeSO_4 + K_2Cr_2O_7 + H_2SO_4 \rightarrow Fe_2(SO_4)_3 + Cr_2(SO_4)_3 + K_2SO_4 + H_2O$$

Since this equation is a little more complex than those we have studied previously, we will solve it carefully and slowly to see why we take each step. First we will list the oxidation numbers giving the whole (SO_4) ion a charge of -2 instead of giving the separate oxidation number of the S and O because we see that the SO_4 ions occur unchanged on both sides of the equation. So we have

$$\overset{+2\ -2}{Fe\ SO_4} + \overset{2(+1)\ 2(+6)\ 7(-2)}{K_2\ \ Cr_2\ \ O_7} + \overset{2(+1)\ -2}{H_2\ \ SO_4} \rightarrow \overset{2(+3)\ 3(-2)}{Fe_2\ \ (SO_4)_3} + \overset{2(+3)\ 3(-2)}{Cr_2\ \ (SO_4)_3} + \overset{2(+1)\ -2}{K_2\ \ SO_4} + \overset{2(+1)-2}{H_2\ \ O}$$

We note that the Fe is changing from $+2$ to $+3$ which is an increase of 1 and the Cr is changing from $+6$ to $+3$ which is a decrease of 3 in oxidation number for each Cr or a total decrease of 6 in oxidation number for the 2 Cr's, or

$$\overset{+2}{FeSO_4} + \overset{2(+6)}{K_2Cr_2O_7} + H_2SO_4 \rightarrow \overset{2(+3)}{Fe_2(SO_4)_3} + \overset{2(+3)}{Cr_2(SO_4)_3} + K_2SO_4 + H_2O$$
$$\uparrow 1 \qquad \downarrow 6$$

The LCM for an increase of 1 and a decrease of 6 in oxidation number is 6. Therefore, an increase of 1 (by the Fe) goes into the LCM 6 times so we place a 6 in front of the $FeSO_4$; a decrease of 6 in oxidation number (by the Cr's in $K_2Cr_2O_7$) goes into the LCM once so we place a 1 (understood) in front of the $K_2Cr_2O_7$, or

$$6\ FeSO_4 + K_2Cr_2O_7 + (\ )\ H_2SO_4 \rightarrow (\ )\ Fe_2(SO_4)_3 + (\ )\ Cr_2(SO_4)_3 + (\ )\ K_2SO_4 + (\ )\ H_2O$$

Now to complete the rest of the equation. Since there are 6 Fe's on the left side there must be 6 on the right and we get these by placing a 3 in front of the $Fe_2(SO_4)_3$ (3 Fe_2 = 6 Fe). Since there are 2 K's on the left side of the equation, there must be 2 on the right and we get these by taking 1 K_2SO_4. Since there are 2 Cr's on the left side, there must be 2 Cr's on the right and so we place a 1 in front of the $Cr_2(SO_4)_3$, or

$$6\ FeSO_4 + K_2Cr_2O_7 + (\ )\ H_2SO_4 \rightarrow 3\ Fe_2(SO_4)_3 + Cr_2(SO_4)_3 + K_2SO_4 + (\ )\ H_2O$$

Now let us count the SO_4 groups on the right side of the equation. There are 9 in the 3 $Fe_2(SO_4)_3$ plus 3 in the $Cr_2(SO_4)_3$ plus 1 in the K_2SO_4 or a total of 13 SO_4's. We started with 6 SO_4's in the 6 $FeSO_4$ so we need 7 more and we get these by taking 7 H_2SO_4. These 7 H_2's on the left side must give 7 H_2's on the right, so we have 7 H_2O, or

$$6\ FeSO_4 + K_2Cr_2O_7 + 7\ H_2SO_4 \rightarrow 3\ Fe_2(SO_4)_3 + Cr_2(SO_4)_3 + K_2SO_4 + 7\ H_2O$$

To check and see if the equation is correctly balanced, let us add up the O's on each side and see if the totals are the same.

Left Side		Right Side	
6 $FeSO_4$	24 O's	3 $Fe_2(SO_4)_3$	36 O's
1 $K_2Cr_2O_7$	7 O's	$Cr_2(SO_4)_3$	12 O's
7 H_2SO_4	28 O's	K_2SO_4	4 O's
		7 H_2O	7 O's
	59 O's		59 O's

and so we see that the equation is correctly balanced.

WHAT IS THE OXIDIZING AGENT? We know that the oxidizing agent is reduced which means that it decreases in oxidation number. The Cr^{+6} in going from $+6$ to $+3$ decreases in oxidation number so, therefore, it is the oxidizing agent; or we might say that the $K_2Cr_2O_7$ is the oxidizing agent since the Cr^{+6} is a part of this compound and the rest of it does not change in oxidation number. Likewise the reducing agent is the Fe^{+2} or the $FeSO_4$ (which is oxidized).

II. THE ION-ELECTRON METHOD OF BALANCING

The balancing of redox equations by the change in oxidation number usually involves quite a few ions which do not change in oxidation number and which are not really necessary for the process of balancing. The ion-electron method eliminates all the unnecessary ions and retains only those which are essential. First we will give the general rules for balancing the "ionic way" and then we will try a few problems to clarify these rules and to see how to use them.

1. Change the equation into 2 half-reactions.

2. Add H^+'s, OH^-'s or H_2O's to either side of each half-reaction to balance the number of atoms of each type.

3. For each half-reaction, add up the charges on each side and equalize them by adding or subtracting electrons from the left side of the half reaction.

4. Find an LCM (least common multiple) for the loss and gain of electrons in the two half-reactions and equalize the loss with the gain.

5. Add the two half-reactions, cancel the lost and the gained electrons, and the equation should be balanced.

Example No. 16-6. Balance: $Fe^{2+} + MnO_4^- + H^+ \rightarrow Fe^{3+} + Mn^{2+} + H_2O$

For this problem we will go through the rules slowly and carefully to see how each applies. After we become familiar with these rules, we will be able to shorten the procedure greatly by doing several steps at once.

rule 1. First we break the original reaction into 2 half-reactions.

$$Fe^{2+} \rightarrow Fe^{3+}$$

$$MnO_4^- \rightarrow Mn^{2+}$$

Let us take the first half-reaction and complete it and then go on the second half-reaction.

rule 2. There are already the same number of iron atoms on each side of the reaction so no other atoms are necessary.

rule 3. The charge on the left side is +2 and on the right it is +3. To equalize them, subtract 1 electron (1 e^-) from the left side of the equation so that the charge on each side is +3.

$$Fe^{2+} - e^- \rightarrow Fe^{3+}$$

$$+2 - (-1) = +3$$

Now to complete the second half-reaction.

rule 2. $MnO_4^- \rightarrow Mn^{2+}$

There are 4 O's on the left side so there must be 4 O's on the right and we get these by adding 4 H_2O to the right side of the half-reaction.

$$MnO_4^- \rightarrow Mn^{2+} + 4 H_2O$$

There are now 8 H's on the right side so we add 8 H^+'s to the left side to equalize the number of atoms. Thus,

$$MnO_4^- + 8 H^+ \rightarrow Mn^{2+} + 4 H_2O$$

where there are 4 O's, 8 H's and 1 Mn on each side.

rule 3. Add up the charges on each side of the half-reaction.

$$MnO_4^- + 8 H^+ \rightarrow Mn^{2+} + 4 H_2O$$

$$-1 \quad\quad + 8 \quad\quad +2 \quad\quad O \quad or$$

$$+7 \quad\quad\quad\quad\quad +2$$

To equalize a charge of +7 on the left with a charge of +2 on the right, we must add 5 negative charges or 5 electrons to the left side, or

$$MnO_4^- + 8 H^+ + 5 e^- \rightarrow Mn^{2+} + 4 H_2O$$

$$[-1 + 8(+1) + 5(-1) \ = \ +2 \ + 4(0)]$$

which is now a completely balanced half-reaction because it contains the same number of each type of atom on each side and also because the charges are balanced. Thus we have:

$$Fe^{2+} - e^- \rightarrow Fe^{3+}$$

$$MnO_4^- + 8 H^+ + 5 e^- \rightarrow Mn^{2+} + 4 H_2O$$

rule 4. Here we have a loss of 1 electron and a gain of 5 so the LCM is 5. For the first half-reaction, 1 goes into the LCM 5 times (1 from 1 electron lost) so we multiply the whole half-reaction by 5 and get

$$5 Fe^{2+} - 5 e^- \rightarrow 5 Fe^{3+}$$

In the second half-reaction, 5 (the number of electrons gained) goes into the LCM once so we write the half-reaction as is (multiply it by 1) and now the loss is equal to the gain.

rule 5. Add the two half-reactions.

$$5 Fe^{2+} - 5 e^- \rightarrow 5 Fe^{3+}$$

$$\underline{MnO_4^- + 8 H^+ + 5 e^- \rightarrow Mn^{2+} + 4 H_2O}$$

$$5 Fe^{2+} + MnO_4^- + 8 H^+ \rightarrow 5 Fe^{3+} + Mn^{2+} + 4 H_2O$$

To check and see if this equation is correctly balanced we note that there are 5 Fe's, 1 Mn, 4 O's and 8 H's on each side of the equation and that the total charge on each side is the same (+17).

The Fe^{2+} is oxidized and so is the reducing agent.

The MnO_4^- is reduced and so is the oxidizing agent.

Example No. 16-7. Balance: $AsO_3^{3-} + I_2 + OH^- \rightarrow AsO_4^{3-} + I^- + H_2O$

First we will break this up into 2 half-reactions, balance each separately and then combine them into a balanced equation. So we have

$$AsO_3^{3-} \rightarrow AsO_4^{3-} \quad and$$

$$I_2 \rightarrow I^-$$

for the first half-reaction,

$$AsO_3^{3-} \rightarrow AsO_4^{3-}$$

there is 1 As on each side but there are 3 O's on the left side and 4 on the right. We equalize them by adding 1 OH$^-$ to the left side, but if we do so we must add H$^+$ to the right side to equalize the H's, or

$$AsO_3^{3-} + OH^- \rightarrow AsO_4^{3-} + H^+$$

However, there can never be H^+ and OH^- in the same reaction, so if we add another OH^- to each side of the equation, we will remove the H^+ as H_2O. Thus,

$$AsO_3^{3-} + 2\ OH^- \rightarrow AsO_4^{3-} + H_2O$$

(We also know from the original equation that H_2O is one of the products.) This half-reaction is balanced as far as atoms are concerned since it contains 1 As, 5 O's, and 2 H's on each side.

Next, to add up the charges,

$$
\begin{array}{ccccccc}
AsO_3^{3-} & + & 2\ OH^- & \rightarrow & AsO_4^{3-} & + & H_2O \\
-3 & + & 2(-1) & & -3 & & 0 \\
& & -5 & & & -3 &
\end{array}
$$

To equalize a charge of -5 on the left with a charge of -3 on the right, we need 2 positive charges on the left side. Since we can use only electrons (negative charges) we must subtract 2 electrons from the left side, or

$$
\begin{array}{ccccccc}
AsO_3^{3-} & + & 2\ OH^- - 2e^- & \rightarrow & AsO_4^{3-} & + & H_2O \\
-3 & & +2(-1)\ \ -2(-1) & & -3 & & 0 \quad \text{or} \\
& & -3-2+2\ \text{does} = -3 & & &
\end{array}
$$

and now we have the half-reaction completely balanced.

The second half-reaction, $I_2 \rightarrow I^-$ we can balance quite easily. Since there are 2 I's on the left side, there must be 2 I's on the right and we need no other atoms so we write

$$I_2 \rightarrow 2\ I^-$$

The charge on the left side is zero and on the right -2. To equalize these charges we add 2 negative charges (electrons) to the left side, or

$$I_2 + 2\ e^- \rightarrow 2\ I^-$$

which is a balanced half-reaction because it contains the same number of I's on each side and also the same charge on each side. Thus we have

$$AsO_3^{3-} + 2\ OH^- - 2\ e^- \rightarrow AsO_4^{3-} + H_2O$$
$$I_2 + 2\ e^- \rightarrow 2\ I^-$$

We can add these directly since the loss is equal to the gain and so

$$
\begin{array}{l}
AsO_3^{3-} + 2\ OH^- - 2\ e^- \rightarrow AsO_4^{3-} + H_2O \\
I_2 + 2\ e^- \rightarrow 2\ I^- \\
\hline
AsO_3^{3-} + I_2 + 2\ OH^- \rightarrow AsO_4^{3-} + 2\ I^- + H_2O
\end{array}
$$

The oxidizing agent is the I_2 since it gained electrons and thus it is reduced.

The reducing agent is the AsO_3^{3-} since it lost electrons and thus it is oxidized.

Example No. 16-8. Balance: $Mn^{2+} + BiO_3^- + H^+ \rightarrow MnO_4^- + Bi^{3+} + H_2O$

First we break it up into half-reactions,

$$Mn^{2+} \rightarrow MnO_4^-$$

$$BiO_3^- \rightarrow Bi^{3+}$$

In the first half-reaction $Mn^{2+} \rightarrow MnO_4^-$, there is 1 Mn on each side but there are 4 O's on the right side and none on the left. To equalize this we can add 4 OH^- or 4 H_2O to the left side. If we add 4 OH^- we will have the same difficulty we had in Example No. 16-7, namely H^+ and OH^- in the same reaction. We can overcome this difficulty as before by removing the H^+'s as H_2O or we can add the H_2O's directly to the left side. Thus,

$$Mn^{2+} + 4\,H_2O \rightarrow MnO_4^-$$

To equalize the number of atoms we need 8 H's on the right, or

$$Mn^{2+} + 4\,H_2O \rightarrow MnO_4^- + 8\,H^+$$

Next, to add up the charges,

$$Mn^{2+} \quad + \quad 4\,H_2O \quad \rightarrow MnO_4^- \quad + \quad 8\,H^+$$
$$+2 \quad + \quad 4\,(0) \quad\quad -1 \quad\quad\quad 8(+1)$$
$$+2 \quad\quad\quad\quad\quad\quad +7$$

To equalize a charge of +2 with a charge of +7, we subtract 5 negative charges from the left side, or

$$\underline{Mn^{2+} + 4\,H_2O - 5\,e^- \quad \rightarrow \quad MnO_4^- + 8\,H^+}$$
$$+2 \quad +0 \quad\quad -5(-1) \quad = \quad\quad -1 \quad +8$$
$$+7 \;=\; +7$$

For the second half-reaction, $BiO_3^- \rightarrow Bi^{3+}$,

there is 1 Bi on each side but 3 O's on the left and none on the right. To equalize this we add 3 H_2O to the right side, or

$$BiO_3^- \rightarrow Bi^{3+} + 3\,H_2O$$

and now we need 6 H's on the left side to balance those on the right. So,

$$BiO_3^- + 6\,H^+ \rightarrow Bi^{3+} + 3\,H_2O$$

Next, to add up the charges,

$$BiO_3^- \quad + \quad 6H^+ \quad \rightarrow \quad Bi^{3+} \quad + \quad 3\,H_2O$$
$$-1 \quad\quad\quad +6 \quad\quad\quad +3 \quad\quad\quad 0$$
$$+5 \quad\quad\quad\quad\quad\quad +3$$

To equalize a +5 and a +3, we add 2 e^- to the left side, or

$$\underline{BiO_3^- + 6\,H^+ + 2\,e^- \rightarrow Bi^{3+} + 3\,H_2O}$$
$$-1 \quad +6 \quad\quad +2(-1) \;=\; +3$$

Thus we have 2 balanced half-reactions:

$$Mn^{2+} + 4\,H_2O - 5\,e^- \rightarrow MnO_4^- + 8\,H^+$$

$$BiO_3^- + 6\,H^+ + 2\,e^- \rightarrow Bi^{3+} + 3\,H_2O$$

Before we can add these two half-reactions, we must equate the loss of electrons with the gain. The LCM for a loss of 5 and a gain of 2 is 10. Five (the loss in the first half-reaction) goes into the LCM twice so we multiply this half-reaction by 2; two (the gain in the second half-reaction) goes into the LCM 5 times so we multiply this half-reaction by 5 and we get:

$$2\ Mn^{2+}\ +\ 8\ H_2O\ -\ 10\ e^-\ \rightarrow 2\ MnO_4^-\ +\ 16\ H^+$$
$$5\ BiO_3^-\ +\ 30\ H^+\ +\ 10\ e^-\ \rightarrow 5\ Bi^{3+}\ \ \ \ \ +\ 15\ H_2O$$

and adding, $2\ Mn^{2+}\ +\ \ \ 5\ BiO_3^-\ +\ 30\ H^+\ +\ 8\ H_2O \rightarrow 2\ MnO_4^-\ +\ \ 5\ Bi^{3+}\ +\ 15\ H_2O\ +\ 16\ H^+$.

We note that there are H_2O's and H^+'s on each side of the equation. We can remove the excess by subtracting $16\ H^+$ and $8\ H_2O$'s from each side to get

$$2\ Mn^{2+}\ +\ 5\ BiO_3^-\ +\ 14\ H^+ \rightarrow 2\ MnO_4^-\ +\ 5\ Bi^{3+}\ +\ 7\ H_2O\ \text{(balanced)}$$

The Mn^{2+} which lost electrons is oxidized and thus is the reducing agent.

The BiO_3^- which gained electrons is reduced and thus is the oxidizing agent.

Problem Assignment

Balance:

16-1. $Cu + HNO_3 \rightarrow Cu(NO_3)_2 + NO + H_2O$
. . . . $(3\ Cu + 8\ HNO_3 \rightarrow 3\ Cu(NO_3)_2 + 2\ NO + 4\ H_2O)$

16-2. $MnO_2 + HCl \rightarrow MnCl_2 + Cl_2 + H_2O$ . . . . $(MnO_2 + 4\ HCl \rightarrow MnCl_2 + Cl_2 + 2\ H_2O)$

16-3. $Bi(OH)_3 + Na_2SnO_2 \rightarrow Bi + Na_2SnO_3 + H_2O$
. . . . $(2\ Bi(OH)_3 + 3\ Na_2SnO_2 \rightarrow 2\ Bi + 3\ Na_2SnO_3 + 3\ H_2O)$

16-4. $Mn(NO_3)_2 + NaBiO_3 + HNO_3 \rightarrow HMnO_4 + Bi(NO_3)_3 + NaNO_3 + H_2O$
. . . $(2\ Mn(NO_3)_2 + 5\ NaBiO_3 + 16\ HNO_3 \rightarrow 2\ HMnO_4 + 5\ Bi(NO_3)_3 + 5\ NaNO_3 + 7\ H_2O)$

16-5. $As_2O_3 + Cl_2 + H_2O \rightarrow H_3AsO_4 + HCl$ . . . $(As_2O_3 + 2\ Cl_2 + 5\ H_2O \rightarrow 2\ H_3AsO_4 + 4\ HCl)$

16-6. $K_2Cr_2O_7 + H_2S + H_2SO_4 \rightarrow K_2SO_4 + Cr_2(SO_4)_3 + S + H_2O$

16-7. $CrO_3 + HI \rightarrow Cr_2O_3 + I_2 + H_2O$

16-8. $Ag + H_2S + O_2 \rightarrow Ag_2S + H_2O$

16-9. $Bi_2S_3 + HNO_3 \rightarrow Bi(NO_3)_3 + NO + S + H_2O$

16-10. $V_2O_5 + KI + HCl \rightarrow V_2O_4 + KCl + I_2 + H_2O$

16-11. $H_3SbO_3 + I_2 + H_2O \rightarrow H_3SbO_4 + HI$

16-12. $PbS + H_2O_2 \rightarrow PbSO_4 + H_2O$

16-13. $CaC_2O_4 + KMnO_4 + H_2SO_4 \rightarrow CaSO_4 + K_2SO_4 + MnSO_4 + CO_2 + H_2O$

16-14. $Mo_2O_3 + KMnO_4 + H_2SO_4 \rightarrow MoO_3 + MnSO_4 + K_2SO_4 + H_2O$

16-15. $Zr + H_2O \rightarrow ZrO_2 + H_2$

16-16. $HNO_3 + SO_2 \rightarrow H_2SO_4 + NO_2$

16-17. $Sb_2(SO_4)_3 + KMnO_4 + H_2O \rightarrow H_3SbO_4 + K_2SO_4 + MnSO_4 + H_2SO_4$

16-18. $K_2Cr_2O_7 + FeCl_2 + HCl \rightarrow FeCl_3 + CrCl_3 + KCl + H_2O$

16-19. $MnO_2 + H_2SO_4 + H_2C_2O_4 \rightarrow MnSO_4 + CO_2 + H_2O$

16-20. $NaCrO_2 + NaClO + NaOH \rightarrow Na_2CrO_4 + NaCl + H_2O$

16-21. $AsH_3 + AgNO_3 + H_2O \rightarrow H_3AsO_3 + Ag + HNO_3$

16-22. $Mo_{24}O_{37} + KMnO_4 + H_2SO_4 \rightarrow MoO_3 + MnSO_4 + K_2SO_4 + H_2O$

16-23. $SbCl_5 + KI \rightarrow SbCl_3 + KCl + I_2$

16-24. $MnCl_2 + NaOH + Br_2 \rightarrow MnO_2 + NaCl + NaBr + H_2O$

16-25. $KIO_4 + KI + HCl \rightarrow KCl + I_2 + H_2O$

16-26. $KIO_3 + H_2SO_3 + HCl \rightarrow KCl + H_2SO_4 + ICl + H_2O$

16-27. $K_2S_2O_8 + H_2C_2O_4 \rightarrow K_2SO_4 + H_2SO_4 + CO_2$

16-28. $Na_2TeO_3 + NaI + HCl \rightarrow NaCl + Te + I_2 + H_2O$

16-29. $CS_2 + O_2 \rightarrow CO_2 + SO_2$

16-30. $Br_2 + NaOH \rightarrow NaBr + NaBrO_3 + H_2O$

16-31. $C_3H_6 + KMnO_4 + H_2O \rightarrow C_3H_8O_2 + MnO_2 + KOH$

16-32. $U_3O_8 + H_2SO_4 \rightarrow UO_2SO_4 + U(SO_4)_2 + H_2O$

16-33. $C_2H_5OH + K_2Cr_2O_7 + H_2SO_4 \rightarrow HC_2H_3O_2 + Cr_2(SO_4)_3 + K_2SO_4 + H_2O$

16-34. $NaCNS + KMnO_4 + H_2SO_4 \rightarrow K_2SO_4 + MnSO_4 + NaCN + H_2O$

16-35. $CrI_3 + KOH + Cl_2 \rightarrow K_2CrO_4 + KIO_4 + KCl + H_2O$

16-36. $Fe^{2+} + Cr_2O_7^{2-} + H^+ \rightarrow Fe^{3+} + Cr^{3+} + H_2O$
. . . . $(6\ Fe^{2+} + Cr_2O_7^{2-} + 14\ H^+ \rightarrow 6\ Fe^{3+} + 2\ Cr^{3+} + 7\ H_2O)$

16-37. $MnO_2 + Cl^- + H^+ \rightarrow Mn^{2+} + Cl_2 + H_2O$
. . . . $(MnO_2 + 2\ Cl^- + 4\ H^+ \rightarrow Mn^{2+} + Cl_2 + 2\ H_2O)$

16-38. $Fe^{3+} + I^- \rightarrow Fe^{2+} + I_2$ . . . . $(2\ Fe^{3+} + 2\ I^- \rightarrow 2\ Fe^{2+} + I_2)$

16-39. $C_2O_4^{2-} + MnO_4^- + H^+ \rightarrow CO_2 + Mn^{2+} + H_2O$
. . . . $(5\ C_2O_4^{2-} + 2\ MnO_4^- + 16\ H^+ \rightarrow 10\ CO_2 + 2\ Mn^{2+} + 8\ H_2O)$

16-40. $Cu + NO_3^- + H^+ \rightarrow Cu^{2+} + NO + H_2O$
. . . . $(3\ Cu + 2\ NO_3^- + 8\ H^+ \rightarrow 3\ Cu^{2+} + 2\ NO + 4\ H_2O)$

16-41. $Zn + NO_3^- + H^+ \rightarrow Zn^{2+} + NH_4^+ + H_2O$

16-42. $ClO^- + H_2O_2 \rightarrow Cl^- + O_2 + H_2O$

16-43. $BrO_3^- + AsO_3^{3-} \rightarrow Br^- + AsO_4^{3-}$

16-44. $SeO_4^{2-} + Cl^- + H^+ \rightarrow SeO_3^{2-} + Cl_2 + H_2O$

16-45. $C_2H_6O + Cr_2O_7^{2-} + H^+ \rightarrow C_2H_4O + Cr^{3+} + H_2O$

16-46. $MnO_2 + C_2O_4^{2-} + H^+ \rightarrow Mn^{2+} + CO_2 + H_2O$

16-47. $IO_4^- + I^- + H^+ \rightarrow I_2 + H_2O$

16-48. $Cu + NO_3^- + H^+ \rightarrow Cu^{2+} + NO_2 + H_2O$

16-49. $BrO_3^- + NH_3 \rightarrow Br^- + N_2 + H_2O$

16-50. $Cr^{3+} + ClO_3^- + OH^- \rightarrow CrO_4^{2-} + Cl^- + H_2O$

16-51. $Fe^{3+} + HS^- \rightarrow Fe^{2+} + S + H^+$

16-52. $Cr_2O_7^{2-} + S^{2-} + H^+ \rightarrow Cr^{3+} + S + H_2O$

16-53. $MnO_4^- + H_2O_2 + H^+ \rightarrow Mn^{2+} + O_2 + H_2O$

16-54. $I_2 + S_2O_3^{2-} \rightarrow S_4O_6^{2-} + I^-$

16-55. $MnO_4^- + NO_2^- + H^+ \rightarrow Mn^{2+} + NO_3^- + H_2O$

16-56. $C_2H_6O + MnO_4^- + H^+ \rightarrow C_2H_4O_2 + Mn^{2+} + H_2O$

16-57. $OCl^- + I^- \rightarrow IO_3^- + Cl^-$

16-58. $CuS + NO_3^- + H^+ \rightarrow Cu^{2+} + SO_4^{2-} + NO + H_2O$

16-59. $H_2O_2 + Fe^{2+} + H^+ \rightarrow Fe^{3+} + H_2O$

16-60. $Pb_3O_4 + H^+ \rightarrow Pb^{2+} + PbO_2 + H_2O$

Chapter 17

THERMOCHEMISTRY

We are all familiar with the term heat in a qualitative way. We know that we have to heat water to raise its temperature and have to add ice or some similar substance to cool it. However, this does not tell us exactly and quantitatively how much heat must be added in the first case and subtracted in the second.

The SI unit of heat is the joule (J). Larger units are the kilojoule (kJ) and the megajoule (MJ). An older unit of heat energy, the calorie, is still in use in medical and nutritional work for the calculation of energy produced by the metabolism of food.

The *specific heat capacity*, c_s, of a substance is defined as the amount of heat required to change the temperature of 1 gram of a substance 1K (or 1°C). Specific heat capacity has the units J/gK or J/g°C.

The amount of heat required to change the temperature of a given amount of a substance without any change in state is given by the formula

$$q = m \times c_s \times \Delta t$$

where m is the mass in grams, c_s is the specific heat capacity of the substance, and Δt is the change in temperature, either in kelvins or °C. q is the amount of heat in joules.

Example No. 17-1. How much heat would be required to change the temperature of 150 grams of water from 25.0°C to 75.0°C? Specific heat capacity of water = 4.18 J/g°C.

Using the formula $q = m \times c_s \times \Delta t$ where m = 150 g, c_s = 4.18 J/g°C and Δt is 50.0°C.

$$q = m \times c_s \times \Delta t = 150 \text{ g} \times 4.18 \text{ J/g°C} \times 50.0°C = 31350 \text{ J} = 31.4 \text{ kJ}$$

Example No. 17-2. How many joules are required to heat 125 g magnesium from 300K to 650K? Specific heat capacity of magnesium is 1.02 J/gK.

$$q = m \times c_s \times \Delta t = 125 \text{ g} \times 1.02 \text{ J/gK} \times 350 \text{ K} = 44600 \text{ J} = 44.6 \text{ kJ}$$

Example No. 17-3. 10.0 grams of a metal at 70.0°C are placed in 10.0 grams of water at 18.0°C. The final temperature of the mixture is 25.0°C. What is the specific heat capacity of the metal?

First, the heat lost by the metal must have been gained by the water.

The heat gained by the water = $m \times c_s \times \Delta t$ = 10.0 g × 4.18 J/g°C × 7.0°C = 293 J.

Thus, the heat lost by the metal must equal 293 J.

So, for the metal, $q = m \times c_s \times \Delta t$ = 10.0 g × c_s × 45.0°C = 293 J

$$c_s = 0.651 \text{ J/g°C}$$

HEATS OF FUSION AND VAPORIZATION

The amount of heat required to melt 1 mole of a substance at its melting point is called the *heat of fusion*, ΔH_{fus}.

The amount of heat required to vaporize 1 mole of a substance at its boiling point is called the *heat of vaporization*, ΔH_{vap}.

For water, ΔH_{fus} = 6.01 kJ/mol and ΔH_{vap} = 40.7 kJ/mol.

Example No. 17-4. How much heat is required to change 5.00 mol ice at 0°C to water at 0°C?

Using the relationship $q = \text{mol} \times \Delta H_{fus}$

$$q = 5.00 \text{ mol} \times 6.01 \text{ kJ/mol} = 30.1 \text{ kJ}$$

Example No. 17-5. How much energy is required to change 150.0 g water at 100°C to steam at 100°C?

$$q = \text{mol} \times \Delta H_{vap}$$

However, we must first change g of water to moles of water.

$$150.0 \text{ g H}_2\text{O} \times \frac{1 \text{ mol H}_2\text{O}}{18.0 \text{ g H}_2\text{O}} = 8.33 \text{ mol H}_2\text{O}$$

Then, $q = 8.33 \text{ mol} \times 40.7 \text{ kJ/mol} = 339 \text{ kJ}$.

ENERGY RELATIONSHIPS IN CHEMICAL REACTIONS

The equation

$$C_{(graphite)} + O_{2(gas)} \rightarrow CO_{2(gas)} + 393.5 \text{ kJ}$$

states that when one mole of graphite combines with one mole of oxygen gas to yield one mole of carbon dioxide gas, 393 kilojoules are liberated. This reaction may also be written as:

$$C_{(graphite)} + O_{2(gas)} \rightarrow CO_{2(gas)} \qquad \Delta H = -393.5 \text{ kJ}$$

where ΔH is called the *heat of formation.*

Note that the first reaction is exothermic; heat is liberated. Note also that in the second reaction ΔH is negative. That is, a negative value of ΔH indicates an exothermic reaction. (Since heat energy is liberated, the heat content of the system must be less.)

Likewise, the reaction

$$2 \text{ C}_{(graphite)} + H_{2(gas)} \rightarrow C_2H_{2(gas)} - 226.7 \text{ kJ}$$

may be written as

$$2 \text{ C}_{(graphite)} + H_{2(g)} \rightarrow C_2H_{2(g)} \qquad \Delta H = 226.7 \text{ kJ}$$

where the positive value of ΔH indicates an endothermic reaction, one in which energy is absorbed.

The following reaction

$$2 \text{ H}_{2(g)} + O_{2(g)} \rightarrow 2 \text{ H}_2\text{O}_{(l)} + 572 \text{ kJ}$$

indicates that 572 kilojoules are liberated upon the formation of 2 moles of liquid water. Since values of ΔH are usually given per mole, the above reaction is written as:

$$H_{2(g)} + \tfrac{1}{2} O_{2(g)} \rightarrow H_2\text{O}_{(l)} \qquad \Delta H = -285.8 \text{ kJ}$$

where the unit (per mole) is understood and not written.

The reverse of the above reaction may be written as:

$$H_2\text{O}_{(l)} \rightarrow H_{2(g)} + \tfrac{1}{2} O_{2(g)} \qquad \Delta H = 285.8 \text{ kJ}$$

That is, ΔH for the reverse of a reaction is the same as that of the original reaction but opposite in sign.

The equation $CH_{4(g)} + 2\ O_{2(g)} \rightarrow CO_{2(g)} + 2\ H_2O_{(g)}$ $\Delta H = -890$ kJ

indicates that when one mole of CH_4 is completely burned to CO_2 and H_2O 890 kJ are evolved. When a reaction involves the complete combustion of a carbon compound with oxygen, the heat involved is called the *heat of combustion*. When a compound is formed from its elements, the heat involved is called the *heat of formation*.

Example No. 17-6.   How many kJ will be evolved in the formation of SO_2 from 8.50 g $S_{(s)}$ and excess $O_{2(g)}$?

$$S_{(s)} + O_{2(g)} \rightarrow SO_{2(g)} \qquad\qquad\qquad \Delta H = -296.8 \text{ kJ}$$

The equation indicates that 297 kJ are evolved in the formation of one mole of SO_2 from its elements. Thus, changing grams S to moles S to moles SO_2 to kJ, we have:

$$8.50 \text{ g S} \times \frac{1 \text{ mol S}}{32.1 \text{ g S}} \times \frac{1 \text{ mol SO}_2}{1 \text{ mol S}} \times \frac{296.8 \text{ kJ}}{1 \text{ mol SO}_2} = 78.6 \text{ kJ}$$

Example No. 17-7.   How much heat will be liberated in the reaction of 2.00 g P with excess Cl_2 to form PCl_5?

$$P + 5/2\ Cl_2 \rightarrow PCl_5 \qquad\qquad\qquad \Delta H = -374.9 \text{ kJ}$$

First, we must change g P to moles P, then to moles PCl_5, and finally to kJ. Recall that heat of formation, ΔH, is in kJ/mole.

$$2.00 \text{ g P} \times \frac{1 \text{ mol P}}{31.0 \text{ g P}} \times \frac{1 \text{ mol PCl}_5}{1 \text{ mol P}} \times \frac{374.9 \text{ kJ}}{1 \text{ mol PCl}_5} = 24.2 \text{ kJ}$$

LAW OF HESS

Hess' Law states that the heat evolved or absorbed in a given chemical process at constant pressure is always the same whether the reaction takes place in one step or in several steps (Law of Conservation of Energy).

This means that thermochemical equations may be treated algebraically to calculate heats of reaction. Thus they can be added or subtracted, multiplied or divided to obtain the desired reactions.

Example No. 17-8.   Calculate ΔH for the reaction $SO_{2(g)} + 1/2\ O_{2(g)} \rightarrow SO_{3(g)}$.

Given:

$$S_{(s)} + O_{2(g)} \rightarrow SO_{2(g)} \qquad\qquad\qquad \Delta H = -296.8 \text{ kJ}$$

$$S_{(s)} + 3/2\ O_{2(g)} \rightarrow SO_{3(g)} \qquad\qquad\qquad \Delta H = -395.7 \text{ kJ}$$

Note that SO_2 in the first reaction is a product rather than a reactant. To make it a reactant we must reverse the first reaction and also change the sign of ΔH. We want SO_3 as a product, which it already is in the second reaction so we leave that reaction as is.

Thus, if we reverse the first reaction and also reverse the sign of ΔH, and then add it to the second reaction, we have

$$SO_{2(g)} \rightarrow S_{(s)} + O_{2(g)} \qquad \Delta H = +296.8 \text{ kJ}$$

$$S_{(s)} + 3/2\ O_{2(g)} \rightarrow SO_{3(g)} \qquad \Delta H = -395.7 \text{ kJ}$$

$$SO_{2(s)} + 1/2\ O_{2(g)} \rightarrow SO_{3(g)} \qquad \Delta H = -98.9 \text{ kJ}$$

Example No. 17-9. Calculate the ΔH for the production of Fe_2O_3 from iron according to the following reaction.

$$2\ Fe + 3\ CO_2 \rightarrow 3\ CO + Fe_2O_3$$

using the following data:

1. $2\ Fe + 1\frac{1}{2}\ O_2 \rightarrow Fe_2O_3 \qquad \Delta H = -822.2 \text{ kJ}$

2. $CO + \frac{1}{2}\ O_2 \rightarrow CO_2 \qquad \Delta H = -283.0 \text{ kJ}$

Note that in the second equation CO is a reactant and not a product as is needed for the required reaction. So if we reverse the second equation, and also change the sign of ΔH, we have:

$$2\ Fe + 1\frac{1}{2}\ O_2 \rightarrow Fe_2O_3 \qquad \Delta H = -822.2 \text{ kJ}$$

$$CO_2 \rightarrow CO + \frac{1}{2}\ O_2 \qquad \Delta H = +283.0 \text{ kJ}$$

However, we cannot add these two reactions directly because the oxygens will not cancel out. (Note that the reaction desired has no O_2's in it.) If we multiply the second reaction by 3 and then add it to the first, then the O_2's will cancel.

$$2\ Fe + 1\frac{1}{2}\ O_2 \rightarrow Fe_2O_3 \qquad \Delta H = -822.2 \text{ kJ}$$

2nd reaction $\times$ 3 $\qquad 3\ CO_2 \rightarrow 3\ CO + 1\frac{1}{2}\ O_2 \qquad 3[\Delta H = +283.0 \text{ kJ}]$

adding $\qquad 2\ Fe + 3\ CO_2 \rightarrow Fe_2O_3 + 3\ CO \qquad \Delta H = 26.8 \text{ kJ}$

Example No. 17-10. Calculate ΔH for the reaction $2C_{(s)} + 3\ H_{2(g)} \rightarrow C_2H_{6(g)}$.

Given:

$$C_{(s)} + O_{2(g)} \rightarrow CO_{2(g)} \qquad \Delta H = -393.5 \text{ kJ}$$

$$H_{2(g)} + \frac{1}{2}\ O_{2(g)} \rightarrow H_2O_{(l)} \qquad \Delta H = -285.8 \text{ kJ}$$

$$C_2H_{6(g)} + 7/2\ O_{2(g)} \rightarrow 2\ CO_{2(g)} + 3\ H_2O_{(l)} \qquad \Delta H = -1559.7 \text{ kJ}$$

In the desired reaction we want C_2H_6 on the right side, so we reverse the third equation and also change the sign of ΔH

$$C_{(s)} + O_{2(g)} \rightarrow CO_{2(g)} \qquad \Delta H = -393.5 \text{ kJ}$$

$$H_{2(g)} + \frac{1}{2}\ O_{2(g)} \rightarrow H_2O_{(l)} \qquad \Delta H = -285.8 \text{ kJ}$$

$$2\ CO_{2(g)} + 3\ H_2O_{(l)} \rightarrow C_2H_{6(g)} + 7/2\ O_{2(g)} \qquad \Delta H = 1559.7 \text{ kJ}$$

The first reaction indicates one mole of CO_2 being formed and the third reaction requires two moles CO_2, so we must multiply the first reaction by 2 (and also multiply the value of ΔH by 2).

$$2\ C_{(s)} + 2\ O_{2(g)} \rightarrow 2\ CO_{2(g)} \qquad\qquad 2\ (\Delta H = -393.5)\ kJ$$

$$H_{2(g)} + \tfrac{1}{2}\ O_{2(g)} \rightarrow H_2O_{(l)} \qquad\qquad \Delta H = -285.8\ kJ$$

$$2\ CO_{2(g)} + 3\ H_2O_{(l)} \rightarrow C_2H_{6(g)} + 7/2\ O_{2(g)} \qquad\qquad \Delta H = 1559.7\ kJ$$

However, in order to cancel the H_2O's we must multiply the second reaction by 3 (in order to cancel the 3 H_2O's in the third reaction).

$$2\ C_{(s)} + 2\ O_{2(g)} \rightarrow 2\ CO_{2(g)} \qquad\qquad 2\ (\Delta H = -393.5)\ kJ$$

$$3\ H_{2(g)} + 3/2\ O_{2(g)} \rightarrow 3\ H_2O_{(l)} \qquad\qquad 3\ (\Delta H = -285.8)\ kJ$$

$$2\ CO_{2(g)} + 3\ H_2O_{(l)} \rightarrow C_2H_{6(g)} + 7/2\ O_{2(g)} \qquad\qquad \Delta H = 1559.7\ kJ$$

$$2\ C_{(s)} + 3\ H_{2(g)} \rightarrow C_2H_{6(g)} \qquad\qquad \Delta H = -84.7\ kJ$$

Heats of Formation of Compounds

ΔH for any reaction is equal to the difference in the heats of formation of the products and the reactants. Table 17-1 lists the heats of formation of various compounds. By definition, the heat of formation of any element is zero.

Table 17-1. Heats of Formation at 25°C

	kJ/mole		kJ/mole		kJ/mole
$AgCl_{(s)}$	-127.0	$CCl_{4(l)}$	-139.3	$MgCl_{2(s)}$	-641.6
$AgNO_{3(s)}$	-124.4	$CO_{2(g)}$	-393.5	$Mg(OH)_{2(s)}$	-924.7
$Al_2O_{3(s)}$	-1,669.8	$CuO_{(s)}$	-156.1	$NaCl_{(s)}$	-410.9
$BaO_{(s)}$	-553.5	$Cu_2O_{(s)}$	-170.7	$NaOH_{(s)}$	-425.6
$CaCO_{3(s)}$	-1207.1	$Fe_2O_{3(s)}$	-822.16	$NH_{3(g)}$	-46.19
$CaO_{(s)}$	-635.5	$HBr_{(g)}$	-36.40	$NH_4Cl_{(s)}$	-314.4
$Ca(OH)_{2(s)}$	-986.2	$HCl_{(g)}$	-92.31	$NO_{2(g)}$	+33.84
$CaSO_{4(s)}$	-1,434.0	$H_2O_{(g)}$	-241.8	$PbO_{(s)}$	-217.3
$CH_3Cl_{(g)}$	-81.9	$H_2O_{(l)}$	-285.8	$PbO_{2(s)}$	-277
$CH_{4(g)}$	-74.8	$HI_{(g)}$	+26.48	$Pb_3O_{4(s)}$	-734.7
$C_2H_{2(g)}$	+226.7	$H_2SO_{4(l)}$	-814.0	$PCl_{3(g)}$	-287.0
$C_2H_{6(g)}$	-84.68	$KCl_{(s)}$	-435.9	$PCl_{5(g)}$	-374.9
$C_5H_{12(g)}$	-173.1	$KNO_{3(s)}$	-492.70	$ZnO_{(s)}$	-348.0

Example No. 17-11. Calculate ΔH for the reaction:

$$CH_{4(g)} + Cl_{2(g)} \rightarrow CH_3Cl_{(g)} + HCl_{(g)}$$

using the values for heats of formation in Table 17-1.

For any given reaction, $\Delta H = \Delta H_f$ products $- \Delta H_f$ reactants

$$\Delta H = \Delta H_f\, CH_3Cl_{(g)} + \Delta H_f\, HCl_{(g)} - \Delta H_f\, CH_{4(g)} - \Delta H_f\, Cl_{2(g)}$$

$$= -81.9 \qquad + (-92.3) \qquad - (-74.8) \qquad - 0$$

$$= -99.4 \text{ kJ}$$

Example No. 17-12. Calculate ΔH for the reaction:

$$C_5H_{12(l)} + 8\, O_{2(g)} \rightarrow 5\, CO_{2(g)} + 6\, H_2O_{(l)}$$

using the table of heats of formation.

From the table we find the following values of ΔH_f: C_5H_{12}, -173.1; $O_{2(g)}$, 0; $CO_{2(g)}$, -393.5; $H_2O_{(l)}$, -285.8

Note that the values of ΔH_f are in kJ/mol. If we have more than one mole of compound we must multiply the value of ΔH_f by that number of moles.

$$\Delta H_f = \Delta H_f \text{ products} - \Delta H_f \text{ reactants}$$

$$= 5\Delta H_f\, CO_{2(g)} + 6\Delta H_f\, H_2O_{(l)} - \Delta H_f\, C_5H_{12(l)} - 8\Delta H_f\, O_{2(g)}$$

$$= 5(-393.5) \qquad + 6(-285.8) \qquad -(-173.1) \qquad -8(0)$$

$$= -3509.2 \text{ kJ}$$

Example No. 17-13. Calculate the heat of formation for $CaCl_2 \cdot 2\, H_2O_{(s)}$ from the following reaction and the table of heats of formation.

$$CaCO_{3(s)} + 2\, HCl_{(g)} + H_2O_{(l)} \rightarrow CaCl_2 \cdot 2\, H_2O_{(s)} + CO_{2(g)} \qquad \Delta H = -122 \text{ kJ}$$

Using $\Delta H = \Delta H_f$ products $- \Delta H_f$ reactants, we have

$$\Delta H = \Delta H_f\, CaCl_2 \cdot 2H_2O_{(s)} + \Delta H_f\, CO_{2(g)} - \Delta H_f\, H_2O_{(l)} - \Delta H_f\, CaCO_{3(s)} - 2\Delta H_f\, HCl_{(g)}$$

$$-122 \text{ kJ} = \Delta H_f\, CaCl_2 \cdot 2\, H_2O_{(s)} + (-393.5) \qquad - (-285.8) \qquad - (-1207.1) \qquad - 2(-92.3)$$

$$\Delta H_f\, CaCl_2 \cdot 2\, H_2O_{(s)} = -1406 \text{ kJ}$$

Heats of Bond Formation

Energy is required to break bonds. That is, bond breaking is an endothermic reaction. Conversely, bond formation is an exothermic process. Table 17-2 lists the heats of bond formation in kJ/mol.

Table 17-2—Heats of Bond Formation at 25°C

	kJ/mol		kJ/mol		kJ/mol
Br-Br	-193	Cl-Cl	-242	H-I	-299
C-Br	-276	F-F	-155	H-O	-463
C-C	-348	H-Br	-366	I-I	-151
C-I	-240	H-F	-567	H-Cl	-431
C-O	-358	H-H	-436	C-Cl	-328
C-H	-413				

Example No. 17-14. Calculate ΔH for the following reaction using heats of bond formation.

$$Br_{2(l)} + 2\ HI_{(g)} \rightarrow 2\ HBr_{(g)} + I_{2(s)}$$

We can use the relationship that ΔH for the reaction is equal to the total heats of bond formation of the products minus the total heats of bond formation of the reactants. We see that in the product 2 HBr, two H-Br bonds are being formed, one for each mole. Likewise, one I-I bond is being formed. From the reactants we see that one Br-Br bond is being broken as are two H-I bonds. Thus,

$$\Delta H = 2\Delta H_f\ (\text{H-Br}) + \Delta H_f\ (\text{I-I}) - \Delta H_f\ (\text{Br-Br}) - 2\Delta H_f\ (\text{H-I})$$

$$= 2(-366) \quad + (-151) \quad - (-193) \quad -2(-299)$$

$$= -92\ kJ$$

Problem Assignment

17-1.   How much heat is required to change 50.0 g Aluminum from 20.0°C to 145°C? c_s for Al = 0.89 J/g°C. . . . . (5.6 kJ)

17-2.   How much heat is required to change the temperature of 35.0 g iron from 400K to 750K? c_s for Fe = 0.449 J/gK. . . . . (5.50 kJ)

17-3.   888 J are used to change the temperature of 12.5 g C from 27.0°C to 127°C. What is the specific heat capacity of carbon? . . . . (0.710 J/g°C)

17-4.   How much heat is required to change 6.75 moles of water to steam at the boiling point?   (275 kJ)

17-5.   How much heat is required to change 200 g CCl_4 from liquid to gas at the boiling point? Molar heat of vaporization of CCl_4 = 30.0 kJ/mol. . . . . (39.0 kJ)

17-6.   How much heat is needed to change 67.4 g SiO_2 from 12.8°C to 137°C? c_s = 0.739 J/g°C.

17-7.   How many kilojoules are needed to change the temperature of 400 mL Hg from 300K to 600K? Density Hg = 13.6 g/mL; c_s = 0.140 J/gK.

17-8.   How much energy is required to change 375 g NH_3 from liquid to gas at the boiling point? ΔH_{vap} = 23.4 kJ/mol.

17-9. How many kilojoules are needed to change 2.75 mol NaCl from solid to liquid at the melting point? ΔH_{fus} = 25.8 kJ/mol.

17-10. 40.0 g of a substance at 150°C are placed in 25.0 g H_2O at 31.0°C. Final temperature of the mixture is 35.5°C. What is the specific heat capacity of the substance? c_s for water = 4.18 J/g°C. . . . . (0.102 J/g°C)

17-11. 300 g of a substance at 400 K are placed in 80.0 g H_2O at 300 K. Final temperature of the mixture is 311 K. What is the specific heat capacity of the substance?

17-12. How many kilojoules would be liberated upon the combustion of 12.5 g of ethylene gas (C_2H_4)?

$$C_2H_4 + 3\ O_2 \rightarrow 2\ CO_2 + 2\ H_2O \qquad\qquad \Delta H = -1388 \text{ kJ/mol } C_2H_4$$

17-13. How many kilojoules would be liberated upon the combustion of 5.60 liters of ethylene at STP? (See Problem 17-12) . . . . (347 kJ)

17-14. How many kilojoules would be absorbed in the combination of 15.5 g of nitrogen with sufficient oxygen to form dinitrogen oxide (N_2O)?

$$N_2 + 1/2\ O_2 \rightarrow N_2O \qquad\qquad \Delta H = 81.6 \text{ kJ}$$

17-15. The heat of formation of NaCl is 411 kJ. How much heat will be liberated upon the reaction of 200 g Na with excess chlorine?

17-16. When 0.400 g of nickel are burned in oxygen, 1636 J are liberated. Calculate the ΔH of NiO. . . . . (-240 kJ/mol)

17-17. A piece of aluminum was burned in oxygen to form the oxide. The heat liberated was sufficient to raise the temperature of 1000 g of water from 15.5° to 22.5°C. What was the mass of the piece of aluminum?

$$2\ Al + 3/2\ O_2 \rightarrow Al_2O_3 \qquad\qquad \Delta H = -1670 \text{ kJ}$$

17-18. A 25.0 g piece of iron was burned in oxygen to form ferric oxide, Fe_2O_3. The heat was sufficient to raise the temperature of some water from 20.0° to 30.0°C. What amount of water was present?

$$2\ Fe + 3/2\ O_2 \rightarrow Fe_2O_3 \qquad\qquad \Delta H = -822.16 \text{ kJ}$$

17-19. The heat of formation of acetylene ($2C + H_2 \rightarrow C_2H_2$) is 227 kJ and that of ethylene ($2C + 2H_2 \rightarrow C_2H_4$) is 52.3 kJ. Calculate the heat of reaction for the process $C_2H_2 + H_2 \rightarrow C_2H_4$. . . . . (-175 kJ)

17-20. The heat of formation of acetylene ($2C + H_2 \rightarrow C_2H_2$) is 227 kJ and that of ethane ($2C + 3H_2 \rightarrow C_2H_6$) is -84.7 kJ. Calculate the heat of reaction for the process $C_2H_2 + 2\ H_2 \rightarrow C_2H_6$.

17-21. The heat of formation of ferrous oxide (Fe + 1/2 O_2 → FeO) is -272 kJ and that of ferric oxide (2 Fe + 3/2 O_2 → Fe_2O_3) is -824 kJ/mol. Calculate the ΔH for the process $2\ FeO + 1/2\ O_2 \rightarrow Fe_2O_3$.

17-22. When 1.000 g of magnesium are burned in oxygen, 24.74 kJ are produced. What is ΔH for the reaction Mg + 1/2 O_2 → MgO?

17-23. How many kilojoules would be liberated in burning 4.60 g P?
$$2P + 5/2\ O_2 \rightarrow P_2O_5 \qquad\qquad \Delta H = -1492 \text{ kJ.}$$

17-24. How many grams of sulfur would have to be burned in order to change 1500 g of water at 100°C to steam at 100°C? ΔH_{vap} = 40.7 kJ/mol.

$$S_{(solid)} + O_{2(gas)} \rightarrow SO_{2(gas)} \qquad \Delta H = -296 \text{ kJ}$$

17-25. Calculate ΔH for the reaction:

$$3 C_{(s)} + 4 H_{2(g)} \rightarrow C_3H_{8(g)}$$

given:

$$C_3H_{8(g)} + 5 O_{2(g)} \rightarrow 3 CO_{2(g)} + 4 H_2O_{(l)} \qquad \Delta H = -2218 \text{ kJ}$$

$$C_{(s)} + O_{2(g)} \rightarrow CO_{2(g)} \qquad \Delta H = -393 \text{ kJ}$$

$$H_{2(g)} + \tfrac{1}{2} O_{2(g)} \rightarrow H_2O_{(l)} \qquad \Delta H = -286 \text{ kJ}$$

17-26. Calculate ΔH for the reaction:

$$2 C_{(s)} + 2 H_{2(g)} + O_{2(g)} \rightarrow CH_3COOH_{(l)}$$

given:

$$CH_3COOH_{(l)} + 2 O_{2(g)} \rightarrow 2 CO_{2(g)} + 2 H_2O \qquad \Delta H = -871 \text{ kJ}$$

$$H_{2(g)} + \tfrac{1}{2}O_{2(g)} \rightarrow H_2O_{(l)} \qquad \Delta H = -286 \text{ kJ}$$

$$C_{(s)} + O_{2(g)} \rightarrow CO_{2(g)} \qquad \Delta H = -393 \text{ kJ}$$

Calculate ΔH in kJ for the following reactions using heats of formation from Table 17-1 (give answers in whole numbers).

17-27. $2 C_2H_{6(g)} + 7 O_{2(g)} \rightarrow 4 CO_{2(g)} + 6 H_2O_{(g)}$ .... (-2855 kJ)

17-28. $Mg(OH)_{2(s)} + 2 HCl_{(g)} \rightarrow MgCl_{2(s)} + 2 H_2O_{(l)}$ .... (-104 kJ)

17-29. $CH_{4(g)} + 4 Cl_{2(g)} \rightarrow CCl_{4(g)} + 4 HCl_{(g)}$ .... (434 kJ)

17-30. $CaCO_{3(s)} \rightarrow CaO_{(s)} + CO_{2(g)}$

17-31. $NH_{3(g)} + HCl_{(g)} \rightarrow NH_4Cl_{(s)}$

17-32. $H_2SO_{4(l)} + CaO_{(s)} \rightarrow CaSO_{4(s)} + H_2O_{(l)}$

17-33. $2 PbO_{(s)} + PbO_{2(s)} \rightarrow Pb_3O_{4(s)}$

Calculate ΔH in kJ for the following problems using heats of bond formation in Table 17-2.

17-34. $H_2 + Br_2 \rightarrow 2 HBr$ .... (-103 kJ)

17-35. $CH_4 + 2 Cl_2 \rightarrow CH_2Cl_2 + 2 HCl$ .... (-210 kJ)

17-36. $F_2 + 2\,HBr \rightarrow 2HF + Br_2$ . . . . (−440 kJ)

17-37. $Cl_2 + 2\,HF \rightarrow 2\,HCl + F_2$

17-38. $CH_4 + 4\,Cl_2 \rightarrow CCl_4 + 4\,HCl$

17-39. $I_2 + 2\,HF \rightarrow F_2 + 2\,HI$

Chapter 18

SOLUTIONS

I. DEFINITIONS

A solution is a homogeneous mixture of two or more substances. A solution, even though it is homogeneous, is not a compound but a mixture, because its composition is variable. For instance, salt dissolved in water makes a salt solution. We can add more salt and get a stronger salt solution or add more water and get a weaker salt solution, but all of these are still salt solutions.

Ordinarily we think of solutions as being liquid, but there are also other types. Air is a solution and is an example of gases dissolved in gases. We also can have solids dissolved in liquids, etc. Solid solutions are too complex to discuss here and gaseous solutions are not too important, because each gas behaves as if it alone were present, so we will devote our time to the discussion of liquid solutions.

Each solution must contain at least two substances. The material that dissolves the other is called the *solvent* and the material that it dissolves is called the *solute*. In the salt solution mentioned above, salt is the solute and water the solvent. An aqueous solution is one in which water is the solvent.

II. METHODS OF EXPRESSING CONCENTRATION

a. Unsaturated, saturated, and supersaturated

If we dissolve a lump of sugar in a cup of coffee, we have a solution which is *unsaturated* because it can still dissolve more sugar. If we keep on adding sugar until no more dissolves, we will obtain a *saturated* solution. If we raise the temperature, we find that we can dissolve still more sugar in the same amount of coffee. Then, if we carefully cool the coffee back to the original temperature, we will find that it will still hold all of the sugar that was dissolved in it previously. This is an example of *supersaturated* solution. If we introduce a crystal of sugar into this supersaturated solution, the excess sugar that we added will crystallize out and a saturated solution will remain. Thus we can define the terms:

unsaturated solution   one which does not contain all of the solute it can at that temperature.

saturated solution   one which holds all of the solute it can dissolve at that temperature. It is in equilibrium with undissolved solute.

supersaturated solution   one which holds more solute than it would ordinarily hold at that temperature.

If we were given a solution and asked to determine whether it was unsaturated, saturated, or supersaturated, we could find out very simply by dropping a crystal of the solute into the solution. If the crystal dissolved, then the solution was unsaturated. If the crystal did not dissolve but remained at the bottom of the container, the solution was saturated. If the crystal caused more crystals to appear when it was added to the solution, then the solution was supersaturated.

It is not very convenient to express concentrations using the above terms, since the solubilities of all substances are different and since we can prepare countless unsaturated solutions of each. Therefore, we need some more exact methods for expressing the concentrations of solutions regardless of the identity of the solute or solvent.

b. Percent composition

Sometimes the solutions are identified by the percent by weight of the solute in a solution of a given density. For example, the label on a bottle of concentrated nitric acid, HNO_3, will read

nitric acid, density 1.5 g/mL, 68% by weight HNO_3

Example No. 18-1. What weight of H_2SO_4 is present in each liter of concentrated sulfuric acid whose density is 1.86 g/mL and which contains 98.0% by weight H_2SO_4?

$$\text{Density} = \frac{\text{Mass}}{\text{Volume}} \text{ and so } 1.86 \text{ g/mL} = \frac{\text{Mass}}{1 \text{ liter or } 1000 \text{ mL}}$$

$$\text{Mass} = 1.86 \frac{g}{mL} \times 1000 \text{ mL} = 1860 \text{ g}$$

Of this, 98.0% is H_2SO_4, so 98.0% of 1860 g or 0.980×1860 g = 1820 g of H_2SO_4 present per liter of solution.

c. Molar solutions

A molar solution is one which contains one mole of solute per liter of solution. Thus, if we take 1 mole of NaCl or 23 g + 35.5 g = 58.5 g and dissolve it in some water and then bring the volume up to 1 liter, we will have a one molar or 1 *M* solution of NaCl in water. A few examples will illustrate this definition better than any other type of explanation.

Example No. 18-2. How many grams of KCl are required to make 1.00 liter of a 2.00 *M* solution?

Since we want 1.00 L of 2.00 *M* KCl and since molarity means moles/liter, we need

$$1.00 \text{ L} \times \frac{2.00 \text{ mol KCl}}{L} \text{ or } 2.00 \text{ mol KCl}$$

Then since 1 mole KCl has a mass of 74.6 g, $2.00 \text{ mol KCl} \times \frac{74.6 \text{ g KCl}}{1 \text{ mol KCl}} = 149 \text{ g KCl}.$

Example No. 18-3. How many grams $MgCl_2$ are required to make 5.00 liters of a 3.00 *M* solution?

Changing molarity to moles/liter and noting that 1 mole $MgCl_2$ has a mass of 95.3 g,

$$5.00 \text{ L} \times 3.00 \text{ } M \text{ } MgCl_2 = 5.00 \text{ L} \times \frac{3.00 \text{ mol } MgCl_2}{L} = 15.0 \text{ mol } MgCl_2$$

$$15.0 \text{ mol } MgCl_2 \times \frac{95.3 \text{ g } MgCl_2}{1 \text{ mol } MgCl_2} = 1.43 \times 10^3 \text{ g } MgCl_2$$

Note that all of the conversion factors could have been placed in one equation:

$$5.00 \text{ L} \times \frac{3.00 \text{ mol } MgCl_2}{L} \times \frac{95.3 \text{ g } MgCl_2}{1 \text{ mol } MgCl_2} = 1.43 \times 10^3 \text{ g } MgCl_2$$

Example No. 18-4. How many grams $BaCl_2$ are required to make 2.00 liters of a 0.400 *M* solution?

Using the above method,
$$2.00 \text{ L} \times \frac{0.400 \text{ mol } BaCl_2}{L} \times \frac{208 \text{ g } BaCl_2}{1 \text{ mol } BaCl_2} = 166 \text{ g } BaCl_2$$

Example No. 18-5. How many grams $FeSO_4$ are needed to make 200 mL of a 0.250 *M* solution?

Since the volume is given in mL and since molarity is expressed in the units moles per liter, we must first change the volume to liters. We also need the molar mass of $FeSO_4$ (1 mole $FeSO_4$ has a mass of 152 g).

$$200 \text{ mL} \times \frac{1 \text{ liter}}{1000 \text{ mL}} \times \frac{0.250 \text{ mol } FeSO_4}{L} \times \frac{152 \text{ g } FeSO_4}{1 \text{ mol } FeSO_4} = 7.60 \text{ g } FeSO_4$$

Next we can work the same type of problem in reverse; that is, given the mass of the solute and the volume of the solution, find the molarity.

Example No. 18-6. What will be the molarity if 1.6 grams NaOH are dissolved to make 125 mL solution?

We have 1.6 g NaOH per 125 mL solution. To find molarity we need moles per liter. Therefore, we must change g/mL to g/L to moles/L.

$$\frac{1.6 \text{ g NaOH}}{125 \text{ mL}} \times \frac{1000 \text{ mL}}{L} \times \frac{1 \text{ mol NaOH}}{40 \text{ g NaOH}} = \frac{0.32 \text{ mol}}{L} = 0.32 \ M$$

$$\text{g/mL} \quad \rightarrow \quad \text{g/L} \quad \rightarrow \quad \text{mol/L} \quad = \quad \text{molarity} \ = \quad M$$

Example No. 18-7. What will be the molarity if 5.0 grams Na_2CO_3 (molar mass 106 g) are dissolved to make 200 mL solution?

Changing g/mL to g/L to moles/L we have the molarity, or

$$\frac{5.0 \text{ g } Na_2CO_3}{200 \text{ mL}} \times \frac{1000 \text{ mL}}{L} \times \frac{1 \text{ mol } Na_2CO_3}{106 \text{ g } Na_2CO_3} = \frac{0.24 \text{ mol } Na_2CO_3}{L} = 0.24 \ M \ Na_2CO_3$$

d. Titration

Titration is a method in which a measured volume of a solution of known concentration is added to a second solution whose concentration is unknown. The point at which the reaction is complete is shown by the color change of an indicator.

Example No. 18-8. 30.0 mL of 0.100 *M* HCl reacts completely with 50.0 mL of NaOH solution. What is the strength of the NaOH solution?

$$HCl + NaOH \rightarrow NaCl + H_2O$$

first, moles HCl = volume HCl (in liters) × molarity HCl (L × mol/L = mol) so

moles HCl = 30.0 mL × $\dfrac{1 \text{ L}}{1000 \text{ mL}}$ × 0.100 mol/L = 0.00300 mol HCl.

From the balanced equation, 1 mol HCl reacts with 1 mol NaOH, so

0.00300 mol HCl will react with 0.00300 mol NaOH.

Then, mol NaOH = $L_{NaOH} \times M_{NaOH}$

$$0.00300 \text{ mol} = 50.0 \text{ mL} \times \frac{1 \text{ L}}{1000 \text{ mL}} \times M$$

$M = \dfrac{0.00300 \text{ mol}}{0.00500 \text{ L}} = 0.0600$ so strength of NaOH = 0.0600 *M*

Example No. 18-9 35.75 mL KOH react completely with 29.34 mL of 0.125 *M* H_2SO_4. What is the strength of the NaOH solution?

$$2KOH + H_2SO_4 \rightarrow K_2SO_4 + 2 \ H_2O$$

Moles H_2SO_4 = 29.34 mL × $\dfrac{1 \text{ L}}{1000 \text{ mL}}$ × 0.125 mol/L = 0.003668

From the balanced equation,

1 mol H_2SO_4 reacts with 2 mol KOH, so

$$\text{mole KOH} = 0.003668 \text{ mol } H_2SO_4 \times \frac{2 \text{ mol KOH}}{1 \text{ mol } H_2SO_4} = 0.007336$$

Then, mol KOH = $L_{KOH} \times M_{KOH}$

$$0.007336 \text{ mol} = 35.75 \text{ mL} \times \frac{1 \text{ L}}{1000 \text{ mL}} \times M$$

$$M = \frac{0.007336 \text{ mol KOH}}{0.03575 \text{ L}} = 0.2052 \text{ } M \text{ KOH}$$

Example No. 18-10. A 5.000 g sample of $CaCO_3$ dissolves completely in 50.00 mL HCl solution. What is the strength of the acid solution?

$$CaCO_3 + 2 \text{ HCl} \rightarrow CaCl_2 + CO_2 + H_2O$$

$$\text{Moles } CaCO_3 = 5.000 \text{ g } CaCO_3 \times \frac{1 \text{ mol } CaCO_3}{100.0 \text{ g } CaCO_3} = 0.05000 \text{ mol } CaCO_3$$

From the balanced equation, 1 mol $CaCO_3$ reacts with 2 mol HCl, so

$$\text{mol HCl} = 2 \times 0.05000 = 0.1000$$

Then, moles = $L \times M$,

$$0.1000 \text{ mol HCl} = 50.00 \text{ mL} \times \frac{1 \text{ L}}{1000 \text{ mL}} \times M$$

$$M = \frac{0.1000 \text{ mol}}{0.05000 \text{ L}} = 2.000, \text{ so strength of acid} = 2.000 \text{ } M$$

Example No. 18-11. 26.14 mL 0.1000 M $KMnO_4$ solution are required to titrate 31.09 mL of $FeSO_4$ solution. What is the strength of the $FeSO_4$ solution?

$$10 \text{ } FeSO_4 + 2 \text{ } KMnO_4 + 8 \text{ } H_2SO_4 \rightarrow K_2SO_4 + 2 \text{ } MnSO_4 + 5 \text{ } Fe_2(SO_4)_3 + 8 \text{ } H_2O$$

$$\text{Mol } KMnO_4 = 26.14 \text{ mL} \times \frac{1 \text{ L}}{1000 \text{ mL}} \times 0.1000 \text{ mol/L} = 0.002614$$

From the balanced equation, we see that

$$0.002614 \text{ mol } KMnO_4 \times \frac{10 \text{ mol } FeSO_4}{2 \text{ mol } KMnO_4} = 0.01307 \text{ mol } FeSO_4$$

$$0.01307 \text{ mol } FeSO_4 = 31.09 \text{ mL} \times \frac{1 \text{ L}}{1000 \text{ mL}} \times M, \text{ and}$$

$$M = \frac{0.01307 \text{ mol}}{0.03107 \text{ L}} = 0.4204 \text{ } M$$

Example No. 18-12. How much iron would have been required to produce the $FeSO_4$ solution in Example 18-11?

From the above example, mol $FeSO_4$ = 0.01307.

$$0.01307 \text{ mol FeSO}_4 \times \frac{1 \text{ mol Fe}}{1 \text{ mol FeSO}_4} \times \frac{55.85 \text{ g Fe}}{1 \text{ mol Fe}} = 0.7300 \text{ g Fe}$$

Equivalents

Equivalents are a method of expressing the amount of a substance taking part in a chemical reaction.

For a reaction involving an acid or base,

$$\text{one equivalent} = \frac{\text{molar mass of acid or base}}{\text{number of } H^+ \text{ or } OH^- \text{ involved}}$$

One equivalent of an acid neutralizes one equivalent of a base.

For a reaction involving oxidation-reduction,

$$\text{one equivalent} = \frac{\text{molar mass of the reactant}}{\text{number of electrons transferred}}$$

One equivalent of an oxidizing agent reacts with one equivalent of a reducing agent.

The **equivalent mass** of a substance is the number of grams of that substance present in one equivalent.

Example 18-13. What is the equivalent mass of HCl?

$$\text{Equivalent mass HCL} = \frac{\text{molar mass HCl}}{\text{number of } H^+ \text{ or } OH^-} = \frac{36.5 \text{ g}}{1} = 36.5$$

Example 18-14. What is the equivalent mass of $Mg(OH)_2$

$$Mg(OH)_2 \rightarrow Mg^{2+} + 2 OH^-$$

$$\text{Equivalent mass Mg(OH)}_2 = \frac{\text{molar mass Mg(OH)}_2}{\text{number of } H^+ \text{ or } OH^-} = \frac{58 \text{ g}}{2} = 29 \text{ g}$$

Example 18-15. What is the equivalent mass of $KMnO_4$ in the reaction

$$KMnO_4 + H_2C_2O_4 + H_2SO_4 \rightarrow K_2SO_4 + MnSO_4 + CO_2 + H_2O$$

In this reaction the Mn in $KMnO_4$ has an oxidation number of +7

The Mn in $MnSO_4$ has an oxidation number of +2

So, the change if oxidation number is 5

$$\text{Equivalent mass KMnO}_4 = \frac{\text{molar mass KMnO}_4}{\text{number of electrons transferred}} = \frac{158 \text{ g}}{5} = 31.6 \text{ g}$$

The same reaction may be written in ionic form

$$MnO_4^- + 8 \, H^+ + 5 \, e^- \rightarrow Mn^{2+} + 4 \, H_2O$$

again showing a transfer of 5 electrons.

Example 18-16. What is the equivalent mass of H_2SO_4 in the following reactions?

a) $H_2SO_4 \rightarrow 2H^+ + SO_4^{2-}$

b) $H_2SO_4 \rightarrow H^+ + HSO_4^-$

in a) equivalent mass of $H_2SO_4 = \dfrac{98 \text{ g}}{2} = 49 \text{ g}$

in a) equivalent mass of $H_2SO_4 = \dfrac{98 \text{ g}}{1} = 98 \text{ g}$

Equivalents are abbreviated as **Eq**; milliequivalents as **mEq**.

Example 18-17. How many mEq are present in 60 mEq KCl?

$$60 \text{ mEq KCl} = 60 \text{ mEq KCl} \times \frac{1 \text{ Eq}}{1000 \text{ mEq}} = 0.060 \text{ Eq KCl}$$

Example 18-18. How many Eq are present in 5.00 g NaOH?

$$\text{Equivalent mass NaOH} = \frac{\text{molar mass NaOH}}{\text{number of } H^+ \text{ or } OH^-} = \frac{40 \text{ g}}{1} = 40 \text{ g}$$

$$5.00 \text{ g NaOH} = 5.00 \text{ g NaOH} \times \frac{1 \text{ Eq NaOH}}{40.0 \text{ g NaOH}} = 0.125 \text{ Eq NaOH}$$

Example 18-19. How many mEq HNO_3 are present in 3.84 mg HNO_3 whose reaction is

$$3Cu + 8 \, HNO_3 \rightarrow 3 \, Cu(NO_3)_2 + 2NO + 4H_2O$$

First, the change in oxidation number of N is from +5 in HNO_3 to +2 in NO, or a change of 3.

$$1 \text{ Eq } HNO_3 = \frac{\text{molar mass } HNO_3}{\text{number of electrons transferred}} = \frac{63.0 \text{ g } HNO_3}{3} = 21.0 \text{ g}$$

Then, changing mg HNO_3 to g HNO_2 to Eq HNO_2 to mEq HNO_3 we have

$$3.84 \text{ mg } HNO_3 = 3.84 \text{ mg } HNO_3 \times \frac{1 \text{ g } HNO_3}{1000 \text{ mg } HNO_3} \times \frac{1 \text{ Eq } HNO_3}{21.0 \text{ g } HNO_3} \times \frac{1000 \text{mEq } HNO_3}{1 \text{ Eq } HNO_3}$$

or 0.183 mEq HNO_3

Example 18-20. In example No 19, how many mEq Cu will react with the 0.183 mEq HNO_3?

Since 1 mEq of one substance reacts with 1 mEq of another, the number of mEq Cu reacting will be 0.183.

Normality

Normality (abbreviated N) is defined as the number of equivalents of an acid or base present in one liter of solution.

$$\text{Normality} = \frac{\text{number of equivalents of acid or base}}{\text{L of solution}}$$

Example 18-21. What will be the normality of a solution containing 0.750 equivalents of HCl in 2.50 L solution?

$$N = \frac{\text{no. of equivalents}}{\text{L of solution}}$$

$$N = \frac{0.750 \text{ Eq}}{2.50 \text{ L}} = 0.300 \text{ N}$$

Example 18-22. What is the normality of a solution which contains 10.0 g NaOH per 500 mL solution?

Changing g NaOH to equivalents NaOH and also changing mL solution to L solution,

$$\frac{10.0 \text{ g NaOH}}{500 \text{ mL}} \times \frac{1 \text{ Eq NaOH}}{40.0 \text{ g NaOH}} \times \frac{1000 \text{ mL}}{1 \text{ L}} = \frac{0.500 \text{ Eq}}{\text{L}} = 0.500 \text{ N}$$

For titration of acid and bases, the following formula may be used

$$\text{volume}_{acid} \times \text{normality}_{acid} = \text{volume}_{base} \times \text{normality}_{base}$$

Example 18-23. 26.7 mL 0.250 N acid will react with how many mL 0.375 N base?

$$\text{Volume}_{acid} \times \text{normality}_{acid} = \text{volume}_{base} \times \text{normality}_{base}$$

$$26.7 \text{ mL} \times 0.250 \text{ N} = V \times 0.375 \text{ N}$$

$$V = \frac{26.7 \text{ mL} \times 0.250 \text{ N}}{0.375 \text{ N}} = 17.8 \text{ mL}$$

e. Molality

A molal solution contains one mole of solute per *1 kilogram of solvent*. We note that this differs from molarity which is given as moles of solute per liter of solution. Thus, 1 mole of NaCl or 58.5 g per 1 kilogram water will yield a 1 molal solution.

Example No. 18-24.   5.30 grams Na_2CO_3 dissolved in 250 grams water will produce a solution of what molality? (Molar mass Na_2CO_3 = 106 g.)

We have 5.30 g Na_2CO_3/250 g H_2O. For molality we need the units moles/1 kg H_2O. Therefore, we must change g/250 g H_2O to g/kg H_2O to moles/kg H_2O.

$$\frac{5.30 \text{ g } Na_2CO_3}{250 \text{ g } H_2O} \times \frac{1000 \text{ g } H_2O}{1 \text{ kg } H_2O} \times \frac{1 \text{ mol } Na_2CO_3}{106 \text{ g } Na_2CO_3} = 0.200 \frac{\text{mol } Na_2CO_3}{1 \text{ kg } H_2O} = 0.200 \text{ molal}$$

$$\text{g/250g } H_2O \;\rightarrow\; \text{g/1kg } H_2O \;\rightarrow\; \text{mol/1kg } H_2O \;=\; m$$

Example No. 18-25.   What is the molality of a 1.000 molar solution of HNO_3 whose density is 1.032 g/mL?

Since the solution is 1.000 molar we have 1.000 mole HNO_3 per liter or 1,000 mL of solution.

$$D = \frac{M}{V} \text{ and } M = D \times V = 1.032 \text{ g/mL} \times 1000 \text{ mL} = 1032 \text{ g.}$$

Therefore, 1.000 liter of solution weighs 1032 g and since it contains 1.000 mole of HNO_3, it contains 63 g HNO_3. If the solution weighs 1032 g and contains 63 g HNO_3, it must contain 1032 g − 63 g or 969 g H_2O.

Thus we have 1 mol HNO_3 in 969 g H_2O, or

$$\frac{1 \text{ mol } HNO_3}{969 \text{ g } H_2O} \times \frac{1000 \text{ g } H_2O}{1 \text{ kg } H_2O} = 1.033 \frac{\text{mol } HNO_3}{1 \text{ kg } H_2O} = 1.033 \; m$$

Example No. 18-26.   What is the molarity of a 2.00 molal solution of H_2SO_4 whose density is 1.12 g/mL?

Since the solution is 2.00 molal it contains 2.00 molar mass H_2SO_4 in 1 kg H_2O or 2.00 × 98 g = 196 g H_2SO_4 in 1000 g H_2O.

Thus the solution contains 2.00 mol H_2SO_4 in 196 g + 1000 g = 1196 g solution.

Then, the volume of the solution = $\dfrac{M}{D}$ = $\dfrac{1196 \text{ g}}{1.12 \text{ g/mL}}$ = 1068 mL = 1.07 L or we have

$$\frac{2.00 \text{ mol } H_2SO_4}{1.07 \text{ L}} = 1.87 \frac{\text{mol } H_2SO_4}{L} = 187 \text{ molar} = 1.87 \; M$$

III. VAPOR PRESSURE OF SOLUTIONS CONTAINING A NON-VOLATILE NON-IONIZED SOLUTE

The vapor pressure of a solution containing a volatile solvent and non-volatile non-ionized solute depends upon the mole fraction of that solvent. This relationship, called Raoult's Law, may be expressed as

$$p = p_0 \left(\frac{n_1}{n_1 + n_2} \right)$$

where p = vapor pressure of solution.

p_0 = vapor pressure of solvent (at the given temperature)

n_1 = number of moles of solvent

n_2 = number of moles of solute.

A solution whose vapor pressure depends on its concentration, according to the above equation (Raoult's Law), is called an *ideal solution*.

The fraction $\dfrac{n_1}{n_1 + n_2}$ is called the **mole fraction** of the solvent.

The mole fraction of solute may be represented as $\dfrac{n_2}{n_1 + n_2}$.

Example No. 18-27. What is the mole fraction of solvent in a solution containing 200 g glucose, $C_6H_{12}O_6$, in 500 g H_2O?

The number of moles of glucose (molecular mass 180) = 200 g $\times \dfrac{1 \text{ mol}}{180 \text{ g}}$ = 1.11 mol

The number of moles of water = 500 g $\times \dfrac{1 \text{ mol}}{18.0 \text{ g}}$ = 27.8 mol

Therefore, mole fraction of solvent = $\dfrac{27.8 \text{ mol}}{27.8 \text{ mol} + 1.11 \text{ mol}}$ = 0.959

Example No. 18-28. What is the vapor pressure at 20°C of a solution containing 200 g of glucose in 500 g of water ? *VP* of water at 20°C = 17.5 torr.

From the data in Example No. 18-16, the mole fraction of solvent = 0.959. Then using the relationship

$$p = p_0 \left(\frac{n_1}{n_1 + n_2} \right)$$

p = 17.5 torr $\times$ 0.959 = 16.7 torr.

Example No. 18-29. Calculate the vapor pressure at 70.0°C of a solution containing 0.100 mole of non-volatile non-ionized solute in 100 g of alcohol (molecular mass 46.0). The vapor pressure of alcohol at 70.0°C is 542 torr.

The number of moles of solvent = 100 g $\times \dfrac{1 \text{ mol}}{46.0 \text{ g}}$ = 2.17 mol.

Therefore, *VP* of solution = $p_0 \left(\dfrac{n_1}{n_1 + n_2} \right)$ = 542 torr $\left(\dfrac{2.17 \text{ mol}}{2.17 \text{ mol} + 0.100 \text{ mol}} \right)$ = 516 torr

Example No. 18-30. How much glycerin (a non-volatile, non-ionized substance of molecular mass 92.0) should be dissolved in 250 g of water at 40.0°C to cause a 2.00 torr lowering of the vapor pressure. *VP* of water at 40.0°C = 55.3 torr.

Since the vapor pressure is lowered 2.00 mm, the *VP* of the solution is 53.3 mm. Thus

$$53.3 \text{ torr} = 55.3 \text{ torr} \left(\frac{n_1}{n_1 + n_2} \right)$$

The number of moles of solvent = $250 \text{ g} \times \dfrac{1 \text{ mol}}{18.0 \text{ g}}$ = 13.9 mol.

Thus, $53.3 \text{ torr} = 55.3 \text{ torr} \left(\dfrac{13.9 \text{ mol}}{13.9 \text{ mol} + n_2} \right)$

n_2 = 0.518 mol of glycerin.

Changing moles of glycerin to grams

$$0.518 \text{ mol} = 0.518 \text{ mol} \times \frac{92.0 \text{ g}}{1 \text{ mol}} = 47.7 \text{ g}$$

Example No. 18-31. What is the molecular mass of a compound if 73.2 g of it dissolved in 200 g of water at 20.0°C lowers the vapor pressure from 17.54 torr to 17.10 torr?

Using the formula $P = P_0 \left(\dfrac{n_1}{n_1 + n_2} \right)$

where the number of moles of solvent = $200 \text{ g} \times \dfrac{1 \text{ mol}}{18.0 \text{ g}}$ = 11.1 mol.

$$17.10 \text{ torr} = 17.54 \text{ torr} \left(\frac{11.1 \text{ mol}}{11.1 \text{ mol} + n_2} \right)$$

n_2 = 0.286 mol.

Therefore, there are 0.286 moles solute present in 73.2 g of that substance.

Thus, $\dfrac{73.2 \text{ g}}{0.286 \text{ mol}}$ = 256 g/mol so that the molecular mass = 256.

IV. VAPOR PRESSURE OF SOLUTIONS CONTAINING A VOLATILE SOLUTE

In an ideal solution of two volatile liquids, both components follow Raoult's Law over the entire range of concentrations. That is, each liquid exerts a vapor pressure proportional to its concentration (mole fraction). The vapor pressure of the solution is merely the sum of the vapor (partial) pressure of its components.

Thus, for a solution containing n_1 moles of A and n_2 moles of B, the total pressure,

$$P_T = P_A \left(\frac{n_1}{n_1 + n_2} \right) + P_B \left(\frac{n_2}{n_1 + n_2} \right)$$

where P_A and P_B are the vapor pressures of components A and B respectively at the given temperature.

Example No. 18-32. What is the vapor pressure of a solution containing 30 mL benzene (C_6H_6) and 80 mL toluene (C_7H_8) at 20.0°C? Vapor pressure of benzene and toluene at 20.0°C are 75 mm and 22 torr respectively. Densities of benzene and toluene at 20.0°C are 0.88 g/mL and 0.87 g/mL respectively.

First, let us calculate the masses, moles, and mole fractions of the two components.

mass benzene = Vol × Density
= 30 mL × 0.88 g/mL = 26.4 g.

mass toluene = Vol × Density
80 mL × 0.87 g/mL = 69.6 g.

$$\text{moles benzene} = \frac{26.4 \text{ g}}{78 \text{ g/mol}} = 0.339 \text{ mol}$$

$$\text{moles toluene} = \frac{69.6}{92 \text{ g/mol}} = 0.757 \text{ mol}$$

$$\text{mole fraction benzene} = \frac{0.339 \text{ mol}}{0.339 \text{ mol} + 0.757 \text{ mol}} = 0.309$$

$$\text{mole fraction toluene} = \frac{0.757 \text{ mol}}{0.339 \text{ mol} + 0.757 \text{ mol}} = 0.691$$

Vapor pressure exerted by benzene = Vapor pressure of benzene × mole fraction benzene

= 75 torr × 0.309 = 23.2 torr.

Vapor pressure exerted by toluene = Vapor pressure of toluene × mole fraction toluene

= 22 torr × 0.691 = 15.2 torr.

Total vapor pressure = 23.2 torr + 15.2 torr = 38.4 torr.

Example No. 18-33. What is the composition, in mole fractions, of the vapor above the liquid mixture of Example No. 18-21.

Since each gas exerts a pressure proportional to its concentration,

$$\text{mole fraction benzene} = \frac{\text{vapor pressure exerted by benzene}}{\text{total vapor pressure of solution}} = \frac{23.2 \text{ torr}}{38.4 \text{ torr}} = 0.604$$

$$\text{mole fraction toluene} = \frac{\text{vapor pressure exerted by toluene}}{\text{total vapor pressure of solution}} = \frac{15.2 \text{ torr}}{38.4 \text{ torr}} = 0.396$$

Example No. 18-34. If the vapor in Example No. 18-22 were collected, condensed to a liquid and then allowed to stand, what would be the composition of the new vapor?

If the vapor from Example No. 18-22 were condensed to a liquid, it would contain benzene and toluene with mole fractions of 0.604 and 0.396 respectively.

Then vapor pressure exerted by benzene = vapor pressure of benzene × mole fraction benzene
= 75 torr × 0.604 = 45.33 torr.

Vapor pressure exerted by toluene = vapor pressure of toluene × mole fraction
= 22 torr × 0.396 = 8.7 torr.

Total vapor pressure of solution = 45.3 torr + 8.7 torr = 54.0 torr.

Then, in the gas phase,

$$\text{mole fraction benzene} \quad \frac{\text{VP exerted by benzene}}{\text{VP of solution}} = \frac{45.3 \text{ torr}}{54.0 \text{ torr}} = 0.839$$

$$\text{mole fraction toluene} \quad \frac{\text{VP exerted by toluene}}{\text{VP of solution}} = \frac{8.7 \text{ torr}}{54.0 \text{ torr}} = 0.161$$

It should be noted that the original mixture with mole fraction of benzene and toluene of 0.309 and 0.691 respectively have become 0.604 and 0.396 respectively after the first evaporation and 0.839 and 0.161 respectively after the second evaporation. Repetition of such a process will lead to a vapor which is nearly pure benzene (and likewise a liquid which is nearly pure toluene).

Example No. 18-35. A mixture of ethyl alcohol and methyl alcohol at 20°C has a *VP* of 60 torr. If the *VP* of ethyl alcohol and methyl alcohol at 20°C are 44 torr and 89 torr respectively, calculate the mole fraction of each in the liquid mixture and in the vapor in equilibrium with that mixture.

If we assume that the mole fraction of ethyl alcohol = *X*, then the mole fraction of methyl alcohol = 1 - *X*.

Then using $P_T = P_A \left(\dfrac{n_1}{n_1 + n_2} \right) + P_B \left(\dfrac{n_2}{n_1 + n_2} \right)$

60 torr = 44 torr (X) + 89 torr $(1 - X)$

60   = 44 X + 89 $(1 - X)$

60   = 44 X + 89 - 89 X

X = 0.644 = mole fraction of ethyl alcohol in liquid mixture.

$1 - X$ = 0.356 = mole fraction of methyl alcohol in liquid mixture.

the *VP* exerted by the ethyl alcohol = *VP* of ethyl alcohol × mole fraction ethyl alcohol
= 44 torr × 0.644 = 28 torr

the *VP* exerted by the methyl alcohol = *VP* of methyl alcohol × mole fraction methyl alcohol
= 89 torr × 0.356 = 32 torr

Then, in the vapor phase,

$$\text{mole fraction ethyl alcohol} = \frac{VP \text{ exerted by ethyl alcohol}}{\text{Total } VP \text{ of solution}} = \frac{28 \text{ torr}}{60 \text{ torr}} = 0.47$$

$$\text{mole fraction methyl alcohol} = \frac{VP \text{ exerted by methyl alcohol}}{\text{Total } VP \text{ of solution}} = \frac{32 \text{ torr}}{60 \text{ torr}} = 0.53$$

V. COLLIGATIVE PROPERTIES OF SOLUTIONS

Raoult's Law states that "any property of a substance in a solution is shown to an extent proportional to the mole fraction of that substance in the solution." These properties which depend only on the number of molecules or particles present irrespective of kind are called *colligative properties*. The vapor pressure of a solution is a colligative property and determines the boiling point and the freezing point of that solution. Therefore, we can say that the elevation of the boiling point or the depression of the freezing point of a liquid is proportional to the mole fraction of the solute in the solvent.

In dilute solutions, the mole fraction of the solute is nearly the same as the molal concentration of the solution so that we can use the relationship that the elevation of the boiling point and the depression of the freezing point are proportional to the molal concentration of the solution.

A one molal solution of a non-volatile non-ionized solute lowers the freezing point of a water solution 1.86°C and raises the boiling point of a water solution 0.52°C. These constants are true only for water, other solvents having their own set of constants.

Thus, a 1 molal solution of sugar (a non-volatile, non-ionized solute) in water will freeze at -1.86°C and a 2 molal solution of the same will freeze at -3.72°C or 2 × (-1.86°C.) This is true regardless of the amount of solution present since the freezing point (and the boiling point) is not determined or affected by the size of the container nor the amount of liquid present.

Likewise, a 1 molal solution of glycerin (also a non-volatile non-ionized solute) in water will boil at 100.52°C (a rise of 0.52°C); a 0.5 molal solution of the same will boil at 100.26°C (0.5 × 0.52°C rise).

The change in boiling point ΔT_b, or the change in the freezing point, ΔT_f, may be calculated from the formulas

$$\Delta T_b = k_b\, m \qquad \text{and} \qquad \Delta T_f = k_f\, m$$

where k_b is the boiling point constant (0.52°C for water), k_f is the freezing point constant (1.86°C for water) and m is the molality of the solution.

Example No. 18-36. What is the boiling point and the freezing point of a solution containing 3.00 moles sugar in 1 kg of water?

Since the solution contains 3.00 moles of sugar in 1 kg H_2O it is a 3.00 molal solution.

Then, $\Delta T_b = k_b\, m = 0.52°C \times 3.00 = 1.56°C$ and the boiling point = 100°C + 1.56°C = 101.56°C.

Likewise, $\Delta T_f = k_f\, m = 1.86°C \times 3.00 = 5.58°C$ so that the freezing point = 0°C - 5.58°C = -5.58°C.

Example No. 18-37. A solution containing 3.10 grams of glycol (molar mass 62.0 g) per 200 grams water will have what freezing point? What boiling point?

First, we must determine the molality of the solution, from which we can calculate the *FP* and the *BP* (freezing point and the boiling point).

$$\frac{3.10 \text{ g glycol}}{200 \text{ g } H_2O} \times \frac{1000 \text{ g } H_2O}{1 \text{ kg } H_2O} \times \frac{1 \text{ mol glycol}}{62.0 \text{ g glycol}} = 0.250 \frac{\text{mol glycol}}{1 \text{ kg } H_2O} = 0.250\ m$$

Then, $\Delta T_f = k_f\, m = 1.86°C \times 0.250 = 0.465°C$ and *FP* = -0.465°C.

Next, $\Delta T_b = k_b\, m = 0.52°C \times 0.250 = 0.13°C$ and *BP* = 100.13°C.

Example No. 18-38. A solution containing 18.4 grams glycerin per 100 grams of water boils at 101.04°C. What is the molecular mass of glycerin?

The boiling point elevation is 101.04°C - 100°C or 1.04°C. Using $\Delta T_b = k_b\, m$,

$$m = \frac{\Delta T_b}{k_b} = \frac{1.04°C}{0.52°C} = 2.0\ m$$

18.4 g glycerin per 100 g water is equivalent to 184 g glycerin per 1 kg water.

2.0 *m* glycerin is equivalent to 2.0 mol glycerin per 1 kg water.

Thus, 184 g glycerin = 2.0 mol and

92 g glycerin = 1 mol so the molecular mass of glycerin = 92.

Example No. 18-39. What is the molecular mass of sugar if a solution of 102.6 g of it dissolved in 200 g water freezes at -2.79°C?

First, let us calculate the molality of the solution. Using $\Delta T_f = k_f\, m$,

$$m = \frac{\Delta T_f}{k_f} = \frac{2.79°C}{1.86°C} = 1.50\ m$$

We have 102.6 g sugar per 200 g water, or 5×102.6 g = 513 g sugar per kg of water. We also have a 1.50 m solution which contains 1.50 moles sugar per kg of water. So we must have 513 g sugar equivalent to 1.50 moles or 513 g/1.50 moles, which is equal to 342 g/mole. Therefore, the molecular mass of sugar must be 342.

Example No. 18-40. How many grams of urea (molar mass 60.0 g) must be added to 250 grams of water to make a solution which will boil at 101.30°C?

The increase in boiling point is 1.30°C and molality = $\dfrac{\Delta T_b}{k_b} = \dfrac{1.30°C}{0.52°C} = 2.5 \ m$.

A 2.5 molal solution contains 2.5 molar masses of solute per kg of water, so we must have 2.5 molar masses urea or 2.5×60.0 g = 150 g urea present per kg of water.

However, we need the amount of urea per 250 grams of water or ¼ of 1 kg, so we take ¼ of 150 grams. Therefore 37.5 grams of urea are necessary.

Example No. 18-41. Pure naphthalene melts at 80.20°C. If 64 grams of Sulfur (molar mass 256 g) are dissolved in 1500 g naphthalene, the resulting solution melts at 79.03°C. What is the freezing point depression constant for naphthalene?

First, we must calculate the molality of the solution and from this we can find the constant we are looking for.

$$\dfrac{64 \text{ g sulfur}}{1500 \text{ g Naph.}} \times \dfrac{1000 \text{ g Naph.}}{1 \text{ kg Naph.}} \times \dfrac{1 \text{ mol sulfur}}{256 \text{ g sulfur}} = 0.17 \dfrac{\text{mol sulfur}}{1 \text{ kg Naph.}} = 0.17 \ m$$

Then, using $\Delta T_f = k_f m$ where ΔT_f = 80.20°C - 79.03°C or 1.17°C,

$$k_f = \dfrac{\Delta T_f}{m} = \dfrac{1.17°C}{0.17} = 6.9°C$$

VI. SOLUTIONS CONTAINING NON-VOLATILE IONIZED SOLUTES

A strong electrolyte is one which is almost completely dissociated in solution, while a weak electrolyte is only partially dissociated. Since colligative properties depend upon the number of particles (or ions) in solution, the degree of dissociation will very definitely affect those colligative properties.

Consider a 0.1 m solution of AB in water.

If AB is non-ionized, the *BP* of the solution will be 100.052°C.

If AB is 100% ionized, then the *BP* of the solution will be 100.104°C (twice the 0.052°C elevation in boiling point) because there will be 2 ions present (per mole), each one affecting the boiling point.

If AB is partially ionized, the *BP* of the solution will be between 100.052°C and 100.104°C.

The relationship between the *BP* and *FP* of a solution of an electrolyte and its molality may be expressed as

$$\Delta T_b = ik_b m \text{ and } \Delta T_f = ik_f m$$

where i, the van't Hoff factor, is the ratio of the colligative effect produced by an "m" molal concentration of an electrolyte divided by the effect for the same concentration of a non-electrolyte. Thus, i may be given by

$$i = \dfrac{\Delta T_b}{(\Delta T_b)_0} \text{ or } i = \dfrac{\Delta T_f}{(\Delta T_f)_0}$$

where $(\Delta T_b)_0$ and $(\Delta T_f)_0$ are the changes in boiling and freezing point respectively for a non-ionized solution of the same molal concentration.

Example No. 18-42. A 0.010 m solution of $Pb(NO_3)_2$ in water freezes at $-0.049°C$. Calculate the value of the van't Hoff factor.

A 0.010 m solution of a non-electrolyte should have a FP of $-0.0186°C$. ($\Delta T_f = k_f m = 1.86°C \times 0.010$)

$$\text{Therefore, } i = \frac{\Delta T_f}{(\Delta T_f)_0} = \frac{0.049°C}{0.0186°C} = 2.63$$

Example No. 18-43. What will be the BP of the 0.010 m solution of $Pb(NO_3)_2$ in example No. 18-31?

Since $\overline{\Delta T_b} = ik_b m$, $\Delta T_b = 2.63 \times 0.52°C \times 0.010 = 0.0137°C$.

Thus, $BP = 100.0137°C$.

The degree of dissociation of the fraction of one mole of an electrolyte that is dissociated into ions may be given by the equation

$$\alpha = \frac{i-1}{n-1}$$

where α is the degree of dissociation, i is the van't Hoff factor and n is the total number of ions yielded by each molecule of the electrolyte.

Example No. 18-44. Calculate the degree of dissociation of the 0.010 m $Pb(NO_3)_2$ solution given in Example No. 18-32.

Using $\alpha = \frac{i-1}{n-1}$ where $n = 3$ since $Pb(NO_3)_2 \rightarrow Pb^{2+} + 2\ NO_3^-$

$$\alpha = \frac{2.63-1}{3-1} = 0.815$$

That is, 0.010 m $Pb(NO_3)_2$ solution is 81.5% ionized.

Example No. 18-34. A 0.010 m $MgSO_4$ solution is 21% dissociated. Calculate the BP of this solution.

Using $\alpha = \frac{i-1}{n-1}$ where $\alpha = 0.21$ and $n = 2$ for $MgSO_4$,

$$0.21 = \frac{i-1}{2-1} = \frac{i-1}{1} \text{ and } i=1.21.$$

Then $\Delta T_b = ik_b m = 1.21 \times 0.52°C \times 0.010 = 0.0063°C$ and the BP of the solution will be $100.0063°C$.

Problem Assignment

18-1.   What weight of solute is necessary to produce:
   a.   2.50 liters of 2.30 M KCl . . . . (373 g)
   b.   300 mL of 0.500 M MgSO$_4$ . . . . (18.0 g)
   c.   150 mL of 0.010 M Na$_3$PO$_4$ . . . . (0.246 g)
   d.   400 mL of 0.3000 M NH$_4$NO$_3$
   e.   4.0 liters of 3.5 M C$_{12}$H$_{22}$O$_{11}$

18-2.   What is the molarity of each of the following solutions:
   a.   20 g KCl in 500 mL solution . . . . (0.54 M)
   b.   100 g KMnO$_4$ in 250 mL solution . . . . (2.5 M)
   c.   84.0 g NaOH in 500 mL solution . . . . (4.20 M)
   d.   400 g C$_{12}$H$_{22}$O$_{11}$ in 1500 mL solution
   e.   4.00 g NaCl in 20.0 mL solution
   f.   1.96 g H$_2$SO$_4$ in 100 mL solution
   g.   140 mg NaOH in 10.0 mL solution

18-3.   What is the molarity of each of the following solutions:
   a.   sulfuric acid solution, density 1.450 g/mL, 55.0% by weight H$_2$SO$_4$ . . . . (8.14 M)
   b.   nitric acid solution, density 1.160 g/mL 27.0% by weight HNO$_3$ . . . . (4.97 M)
   c.   acetic acid solution, density 1.026 g/mL, 20.0% by weight HC$_2$H$_3$O$_2$
   d.   hydrochloric acid solution, density 1.148 g/mL, 29.1% by weight HCl

18-4.   What weight of solute is present per mL in each of the following:
   a.   sodium hydroxide solution, density 1.09 g/mL, 9.00% by weight NaOH . . . . . . (0.0981 g)
   b.   sulfuric acid solution, density 1.86 g/mL, 98.0% by weight H$_2$SO$_4$
   c.   nitric acid solution, density 1.65 g/mL, 69.4% by weight HNO$_3$
   d.   5.00% by weight C$_6$H$_{12}$O$_6$ solution, density 1.04 g/mL
   e.   0.150 M NaOH solution . . . . (0.00600 g)
   f.   2.50 M C$_3$H$_8$O$_3$ solution

18-5.   48.74 mL 0.1500 M HCl react with 46.50 mL of what strength KOH?

   KOH + HCl → KCl + H$_2$O . . . . (0.1572 M)

18-6.   30.00 mL 0.0650 M Ca(OH)$_2$ react with 35.00 mL HNO$_3$ of what strength?

   Ca(OH)$_2$ + 2 HNO$_3$ → Ca(NO$_3$)$_2$ + 2 H$_2$O . . . . (0.111 M)

18-7.   300 mL 0.160 M HCl are needed to dissolve how many g SrCO$_3$?

   SrCO$_3$ + 2 HCl → SrCl$_2$ + CO$_2$ + H$_2$O . . . . (3.54 g)

18-8.   24.62 mL 0.09250 M H$_2$C$_2$O$_4$ react with 22.59 mL KMnO$_4$. What is the strength of the KMnO$_4$?

   2 KMnO$_4$ + 5 H$_2$C$_2$O$_4$ + 3 H$_2$SO$_4$ → K$_2$SO$_4$ + 2 MnSO$_4$ + 10CO$_2$ + 8 H$_2$O . . . . . (0.0403 M)

18-9.   42.7 mL 0.125 M H$_2$SO$_4$ will react with how many mL 0.175 M BaCl$_2$?

   BaCl$_2$ + H$_2$SO$_4$ → BaSO$_4$ + 2 HCl

18-10. 50.00 mL 0.1000 M HCl will react with how many mL 0.0850 M KOH?

$$KOH + HCl \rightarrow KCl + H_2O$$

18-11. 29.87 mL 0.195 M $FeSO_4$ will react with 23.74 mL of what strength $K_2Cr_2O_7$?

$$K_2Cr_2O_7 + 6\ FeSO_4 + 7\ H_2SO_4 \rightarrow K_2SO_4 + Cr_2(SO_4)_3 + 3\ Fe_2(SO_4)_3 + 7\ H_2O$$

18-12. 1.50 g $CaCO_3$ will dissolve in what volume of 0.150 M HCl?

$$CaCO_3 + 2\ HCl \rightarrow CaCl_2 + CO_2 + H_2O$$

18-13. 5.70 g zinc will dissolve in 300 mL of what strength HCl?

$$Zn + 2\ HCl \rightarrow ZnCl_2 + H_2$$

18-14. What is the equivalent mass of $Al(OH)_3$

$$Al(OH)_3 + 3\ HCl \rightarrow AlCl_3 + 3\ H_2O \quad \dots\dots\dots\dots\dots\dots\dots \textbf{(26.06g)}$$

18-15. What is the equivalent mass of H_2SO_4?

$$H_2SO_4 + KOH \rightarrow KHSO_4 + H_2O$$

18-16. What is the equivalent mass of KIO_3?

$$KIO_3 + 2\ H_2SO_3 + 2\ HCl \rightarrow KCl + 2\ H_2SO_4 + ICl + H_2O \dots\dots\dots\dots \textbf{(53.5 g)}$$

18-17. What is the equivalent mass of Cr^{3+}?

$$2\ Cr^{3+} + ClO_3^- + 10\ OH^- \rightarrow 2\ CrO_2^{2-} + Cl^- + 5\ H_2O$$

18-18. How many mEq are present in 4.73 g H_2SO_4?

$$H_2SO_4 \rightarrow 2H^+ + SO_6^{2-} \quad \dots\dots\dots\dots\dots\dots\dots\dots\dots \textbf{(0.0965)}$$

18-19. How many mEq are present in 1.00 g LiOH?

18-20. How many mEq are present in 1.47 g HNO_3?

18-21. How many mEq are present in 500 mg KOH? $\dots\dots\dots\dots\dots\dots\dots\dots\dots$ **(8.91)**

18-22. What is the normality of a solution containing 0.0318 mEq of an acid in
200 mL solution? $\dots\dots\dots\dots\dots\dots\dots\dots\dots\dots\dots\dots$ **(0.159N)**

18-23. What is the normality of a solution containing 3.00 mEq base in 400 mL solution. . . **(0.750N)**

18-24. How many mL 0.308 N H_2SO_4 are required to neutralize 27.9 mL 0.400 N NaOH? **(36.2 mL)**

18-25. 20.4 mL 0.0500 N acid are required to neutralize 38.6 mL of what strength base?

18-26.   What is the molality of a solution containing:
   a.   120 g $C_6H_{12}O_6$ in 500 g H_2O .... (1.33 m)
   b.   45.0 g $CuSO_4$ in 250 g H_2O .... (1.13 m)
   c.   2.50 moles alcohol in 300 g H_2O .... (8.33 m)
   d.   8.65 g naphthalene ($C_{10}H_8$) in 100 g benzene (C_6H_6)
   e.   125 g $C_3H_8O_3$ in 1200 g H_2O
   f.   1.5 moles NaOH in 25 moles H_2O

18-27.   How many grams of solute should be dissolved in 200 g H_2O to prepare:
   a.   2.50 m H_2SO_4 solution .... (49.0 g)
   b.   0.200 m $KMnO_4$ .... (4.74 g)
   c.   0.150 m NaCl solution
   d.   0.350 m C_2H_5OH solution

18-28.   What is the percent by weight of solute in:
   a.   3.00 m $C_6H_{12}O_6$ solution .... (35.1%)
   b.   0.200 m NaCl solution
   c.   1.50 m $AgNO_3$ solution

18-29.   What are the boiling points and the freezing points of the following solutions of non-ionized solutes:
   a.   1.55 g glycol, $C_2H_6O_2$, in 200 g H_2O .... (100.065°C, –0.232°C)
   b.   12.0 g glucose, $C_6H_{12}O_6$, in 400 g H_2O .... (100.087°C, –0.310°C)
   c.   20.0 g glycerin, $C_3H_8O_3$, in 250 g H_2O
   d.   150 g alcohol, C_2H_6O, in 800 g H_2O
   e.   2.50 g naphthalene, ($C_{10}H_8$), in 100 g benzene, C_6H_6.
          FP benzene = 5.50°C, BP benzene = 80.10°C
          K_f = 4.90°C, K_b = 2.53°C
   f.   3.44 g hexane, C_6H_{14}, in 50 g ether,
          BP ether = 34.60°C, FP ether = –116.30°C
          K_f = 1.79°C, K_b = 2.16°C

18-30.   Calculate the molecular mass of the non-ionized solute in each of the following solutions:
   a.   24 g solute in 250 g H_2O freezes at –3.72°C. .... (48)
   b.   45.0 g solute in 250 g H_2O boils at 100.26°C .... (360)
   c.   10.0 g solute in 100 g naphthalene freezes at 75.2°C.
          FP naphthalene = 80.2°C, K_f = 6.9°C .... (138)
   d.   15.0 g solute in 200 g H_2O boils at 100.39°C
   e.   25.0 g solute in 180 g H_2O freezes at –1.40°C
   f.   2.00 g solute in 10.0 g H_2O boils at 101.26°C
   g.   1.00 g solute in 50.0 g CCl_4 boils at 77.80°C.
          BP = 76.80°C, K_b = 5.03°C.
   h.   2.50 g solute in 100 g camphor freezes at 166°C.
          FP = 176°C, K_f = 49.8°C.

18-31.   A solution contains 2.0 moles NaCl and 50 moles H_2O. What is the mole fraction of the solute? of the solvent?

18-32.   A water solution contains 10.0% by weight $MgSO_4$. What is the mole fraction of the solute? What is the molality of the solution?

18-33. A sulfuric acid solution has a density of 1.381 g/mL and contains 48.1% by weight H_2SO_4. Calculate its M, m, and mole fraction of solute.

18-34. How many grams of a non-ionized antifreeze (molecular mass 62) should be poured into an automobile radiator (capacity 20 quarts) to keep the liquid from freezing until $-15°F$? (1 liter = 1.057 quarts). (Assume density of solution to be 1.00 g/mL.)

18-35. Assuming 100% ionization, what will be the FP and BP of the following solutions:
    a.   0.850 m H_2SO_4
    b.   16.2 g LiCl in 200 g H_2O
    c.   1.38 g $NaNO_3$ in 100 g H_2O

18-36. Calculate the boiling point elevation constant from the following data: 10.0 g of a solute (molecular mass 200) dissolved in 250 g of a solvent (BP 61.26°C) boils at 61.89°C. (Assume solute is non-ionized.)

18-37. A non-electrolyte contains 38.7% C, 9.7% H, and 51.6% O. 15.5 g of this solute dissolved in 100 g H_2O boils at 101.30°C. What is the molecular formula of the solute?

18-38. A solution containing 1.60 moles of non-ionized solute freezes at $-9.00°C$. What amount of water is present in the solution?

18-39. Which of the following non-ionized solutions has the lowest freezing point and what is it?
    a.   6.80 g alcohol, C_2H_6O, in 100 g H_2O
    b.   22.3 g glycerin, $C_3H_8O_3$, in 150 g H_2O
    c.   128 g sugar, $C_{12}H_{22}O_{11}$, in 250 g H_2O

18-40. A solution containing 12.4 g of phosphorus in 40.0 g CS_2 boils at 52.15°C. What is the molecular formula of phosphorus?
BP CS_2 = 46.30°C, K_b = 2.34°C.

18-41. 100 mL of 2.00 M H_2SO_4 solution:
    a.   contains how many grams H_2SO_4?
    b.   contains how many moles H_2SO_4?
    c.   can be prepared from what volume of 6.00 M H_2SO_4?
    d.   will react with excess metal to produce how many moles H_2?
    e.   will react with excess metal to produce how many grams H_2?
    f.   will react with excess metal to produce how many molecules H_2?

18-42. What is the vapor pressure of a solution which contains 100 g glycerol ($C_3H_8O_3$) in 450 g H_2O at 15.0°C? VP of water at 15.0°C = 12.8 torr.
        (12.3 torr)

18-43. What is the vapor pressure of a solution which contains 1.50 moles of a non-volatile non-ionized compound in 194 mL of CCl_4 at 23.9°C? Density and vapor pressure of CCl_4 at 23.0°C are 1.59 g/mL and 100 mm respectively.

18-44. How many grams of sucrose ($C_{12}H_{22}O_{11}$) should be dissolved in 252 g of water at 18.0°C to cause the resulting solution to have a vapor pressure of 15.0 torr? VP of water at 18.0°C = 15.4 mm.

18-45.   How many grams of fructose ($C_6H_{12}O_6$) should be dissolved in 180 g water at 31.6°C to lower the vapor pressure 1.00 torr? *VP* of water at 31.6°C = 34.9 torr. . . . . (53.1 g)

18-46.   What is the molecular mass of a compound if 3.00 g of it dissolved in 25.0 g of water at 22.0°C produced a solution of *VP* 20.5 torr? *VP* of water at 22.0°C = 21.1 torr.

18-47.   What is the molecular mass of a substance if 5.00 g of it dissolved in 200 g of CCl_4 lowers the vapor pressure from 86.8 torr to 85.1 torr? . . . . (193)

18-48.   A 0.20 *m* solution of NH_4Cl has a boiling point of 100.189°C. Calculate the value of *i*, the van't Hoff factor. . . . . (1.82)

18-49.   A 0.0050 *m* $K_3Fe(CN)_6$ solution has a freezing point of –0.0326°C. Calculate the value of *i*, the van't Hoff factor.

18-50.   Calculate the freezing point of the NH_4Cl solution to Problem No. 18-48. . . . . . . (–0.677°C)

18-51.   Calculate the boiling point of the solution in Problem No. 18-49.

18-52.   Calculate the degree of dissociation in Problem No. 18-48. . . . . (0.82)

18-53.   Calculate the degree of dissociation in Problem No. 18-49.

18-54.   A 0.10 *m* $CoCl_2$ solution is 87.5% ionized. What is the *BP*? . . . . (100.14°C)

18-55.   A 0.40 *m* K_2SO_4 solution is 52% ionized. What is the freezing point?

18-56.   What is the freezing point of the $CoCl_2$ solution in Problem No. 18-54? . . . . (–0.51°C)

18-57.   A mixture of benzene (C_6H_6) and toluene (C_7H_8) at 30.0°C contains 80.0% by weight benzene. What is the vapor pressure of the solution? *VP* of benzene and toluene at 60.0°C = 118 mm and 36.7 mm respectively. . . . . (104 torr)

18-58.   A mixture of ethyl alcohol (C_2H_6O) and methyl alcohol (CH_4O) at 60.0°C contains 10.0% methyl alcohol. What is the *VP* of the solution? *VP* of ethyl alcohol and methyl alcohol at 60.0°C are 353 torr and 625 torr respectively.

18-59.   What is the composition in mole fractions of the vapor above the liquid in Problem No. 18-57? . . . . (0.938 benzene; 0.062 toluene)

18-60.   What is the composition, in mole fractions, of the vapor above the liquid in Problem No. 18-58?

18-61.   If the vapor in Problem No. 18-57 were collected, condensed and then allowed to stand, what would be the composition, in mole fractions, of the vapor? . . . (0.980 benzene; 0.020 toluene)

18-62.   If the vapor in Problem No. 18-58 were collected, condensed and then allowed to stand, what would be the composition, in mole fractions, of the vapor?

18-63.   A mixture of two liquids A and B at 25.0°C has a vapor pressure of 100 torr. If the vapor pressures of A and B at 25.0°C are 60.0 torr and 120 torr respectively, what is the mole fraction of each in the mixture? . . . . (0.333 A and 0.667 B)

18-64.   A mixture of two liquids X and Y at 100°C has a vapor pressure of 250 torr. If the vapor pressures of X and Y are 140 torr and 300 torr respectively, what is the mole fraction of each in the solution and also in the vapor above that solution?

18-65.   What is the mole fraction of A and B in the vapor above the liquid in Problem No. 18-63? (0.200 A; 0.800 B)

18-66.   A mixture of two volatile liquids R and S at a given temperature has a vapor pressure of 242 torr. If the mole fraction of R in the liquid is 0.720 and is 0.537 in the vapor, calculate the vapor pressures of pure R and S at the given temperature.

Chapter 19

EQUILIBRIUM REACTIONS

Many times when two or more reactants unite to form a certain number of products, these products themselves unite to re-form the original reactants. Reactions of this type are called *reversible reactions*. They are indicated by double arrows ⇌ showing that the reaction may proceed in either direction depending upon the conditions that exist.

If we start with a mixture of N_2 and H_2, at a given temperature and pressure (with a catalyst), we will soon have some NH_3 formed. As more NH_3 is formed, it will begin to decompose into N_2 and H_2 or

$$N_2 + 3\,H_2 \rightleftharpoons 2\,NH_3$$

When the rates of formation and decomposition become equal, a chemical equilibrium exists. This does not mean that all reaction has stopped; it merely means that the rate of decomposition is the same as the rate of formation so that the composition remains constant.

The *Law of Mass Action* states that the rate of a chemical reaction is proportional to the molecular concentration of the reacting substances.

For liquids and dissolved solids, concentrations are usually expressed in moles per liter; for solids, since crystalline structure and solubility definitely fix the concentration, these concentrations are usually ignored; for gases, since concentration in moles per liter is greatly affected by the applied pressure, concentrations are usually expressed in terms of partial pressure of the gases.

$$\text{Partial pressure of a gas} = \frac{\text{number of moles of the gas present}}{\text{total number of moles of gases present}} \times \text{total pressure}$$

EQUILIBRIUM CONSTANT

If, in the reaction $A + B \rightleftharpoons C + D$, we allow the bracket sign [ ] to indicate concentration in moles per liter, then

$$[A] \quad = \quad \text{concentration of A in moles per liter}$$

$$[B] \quad = \quad \text{concentration of B in moles per liter}$$

$$[C] \quad = \quad \text{concentration of C in moles per liter}$$

$$[D] \quad = \quad \text{concentration of D in moles per liter}$$

For the forward reaction, the rate will be proportional to the concentration of the reactants, according to the law of mass action, or

rate forward ($\rightarrow$) = $k_1 \times [A] \times [B]$ where k_1 is a proportionality constant. The rate of the reverse reaction will depend upon the concentration of the products and so will be proportional to their concentrations, or
rate of reverse reaction ($\leftarrow$) = $k_2 \times [C] \times [D]$ where k_2 is another proportionality constant.

At equilibrium, the rate of the forward reaction is equal to the rate of the reverse reaction, so

$$k_1 \times [A] \times [B] = k_2 \times [C] \times [D] \quad \text{whence}$$

$$\frac{[C] \times [D]}{[A] \times [B]} = \frac{k_1}{k_2} = K_c \quad \text{where } K_c \text{ is the concentration equilibrium constant (since the ratio of 2 constants } k_1/k_2 \text{ is itself another constant).}$$

Likewise, for the reaction

$$X + Y + Z \rightleftharpoons P + Q + R + S$$

the concentration equilibrium constant is

$$K_c = \frac{[P] \; [Q] \; [R] \; [S]}{[X] \; [Y] \; [Z]}$$

and for the reaction $4\,A + 3\,C \rightleftharpoons 2\,D + F$,

$$K_c = \frac{[D]^2 \; [F]}{[A]^4 \; [C]^3}$$

since we could have written the reaction as

$$A + A + A + A + C + C + C \rightleftharpoons D + D + F \qquad\qquad (4\,A + 3\,C \rightleftharpoons 2\,D + F)$$

whence $K_c = \dfrac{[D] \; [D] \; [F]}{[A] \; [A] \; [A] \; [A] \; [C] \; [C] \; [C]} = \dfrac{[D]^2 \; [F]}{[A]^4 \; [C]^3}$

Thus we see that the concentration equilibrium constant (K_c) equals the product of the concentrations of the products divided by the product of the concentrations of the reactants, each concentration raised to the power indicated by its coefficient in the equation.

The value of the equilibrium constant is dependent upon temperature only and is unaffected by other variables such as pressure, concentration, presence of a catalyst, etc.

A large value of an equilibrium constant denotes an equilibrium that proceeds far to the right (greater concentration of products) while a low value indicates an equilibrium reaction that does not proceed very far to the right, but rather to the left (greater concentration of reactants).

For gases, if we use the partial pressures to express concentrations instead of moles per liter, we have K_p, the partial pressure equilibrium constant.

Thus, for the reaction $N_2 + 3H_2 \rightleftharpoons 2\,NH_3$

$$K_p = \frac{p^2_{NH_3}}{p_{N_2} \, p^3_{H_2}}$$

where P_{NH_3} is the partial pressure of the NH_3 in the mixture at equilibrium, P_{N_2} is the partial pressure of the N_2, and P_{H_2} is the partial pressure of the H_2, each partial pressure being raised to the power indicated by its coefficient in the equilibrium reaction.

For the general formula

$$a\,X + b\,Y \rightleftharpoons c\,Z + d\,W \qquad\qquad (X,\, Y,\, Z,\, \text{and } W \text{ being gases})$$

$$K_p = \frac{p_Z{}^c \cdot p_W{}^d}{p_X{}^a \cdot p_Y{}^b}$$

LE CHATELIER'S PRINCIPLE

If a stress is applied to a reaction at equilibrium, the equilibrium will be displaced in such a direction as to relieve that stress. Thus, if we apply a stress such as change of pressure, temperature, in concentration, etc., we should be able to predict the results of this stress upon the equilibrium.

A. Effect of Concentration

Let us see what the effect will be upon the equilibrium constant and upon the equilibrium itself for the reaction

$$A + B \rightleftharpoons C + D$$

if we add more of A to the mixture of A, B, C, and D at equilibrium.

The addition of more A will cause a stress which the reaction will tend to oppose by using up more A and so shifting the equilibrium to the right. Since the *equilibrium constant* depends upon temperature only, the addition of more A will have no effect upon it.

Also, since $K_c = \dfrac{[C] \; [D]}{[A] \; [B]}$ when we increase [A], [C] and [D] must also increase to give the same value of K_c and so, as we reasoned above, the equilibrium will shift to the right. The same will hold true if we use K_p.

B. Effect of Pressure

In the reaction

$N_2 + 3 H_2 \rightleftharpoons 2 NH_3$ (since all are gases)

1 volume + 3 volumes $\rightleftharpoons$ 2 volumes, or

4 volumes of gas $\rightleftharpoons$ 2 volumes of gas

Increasing the pressure on the system will set up a stress and the system will tend to oppose this stress by shifting in the direction which tends to lower the pressure. That is, it will tend to go to a smaller volume. Thus, an increase in pressure will cause the equilibrium to shift to the right (smaller volume) and a decrease in pressure will cause a shift to the left. This holds true only for gases in a mixture at equilibrium, solids and liquids being unaffected by pressure changes.

The *equilibrium constant* itself remains unchanged.

Pressure will have no effect upon a reaction at equilibrium when the number of gas molecules of reactant is the same as the number of gas molecules of product. Thus in the reaction

$H_2 + Cl_2 \rightleftharpoons 2 HCl$ (all gases)

2 volumes $\rightleftharpoons$ 2 volumes

pressure will have no effect upon the equilibrium.

C. Effect of Temperature

In the reaction

$4 HCl + O_2 \rightleftharpoons 2 H_2O + 2 Cl_2$ (all gases) $\qquad \Delta H = -59.4$ kJ, heat is liberated when the reaction proceeds to the right and is absorbed when it proceeds to the left.

If we raise the temperature of this reaction mixture at equilibrium, the reaction will tend to go in a direction to relieve this stress. That is, as the temperature is raised, the equilibrium will shift to the left favoring the reaction which tends to absorb this heat. Conversely, as the temperature is lowered, the reaction speeds more to the right for the production of more heat to relieve the new stress.

Since all are gases, the equilibrium constant for this reaction is

$$K_p = \frac{p^2_{H_2O} \cdot p^2_{Cl_2}}{p^4_{HCl} \cdot p_{O_2}}$$

As the temperature is raised, more HCl and O_2 are formed and less H_2O and Cl_2 are left so that K_p decreases. Likewise, lowering the temperature will increase K_p.

NOTE: A change in temperature is the only factor that will affect the value of the equilibrium constant.

D. Effect of a Catalyst

The addition of a catalyst will speed up the forward reaction but will also speed up the reverse reaction equally so that all that is accomplished is that the system reaches equilibrium much sooner. The equilibrium reached is the same as would have been reached if no catalyst had been used. Again we note that temperature is the only thing that affects the equilibrium constant, so naturally a catalyst will have no effect on it.

Example No. 19-1. For the reaction

$$C_2H_5OH + CH_3COOH \rightleftharpoons CH_3COOC_2H_5 + H_2O$$

the concentrations at equilibrium are:

$[C_2H_5OH]$ = 1/3 mole per liter; $[CH_3QOH]$ = 1/3 mole per liter; $[CH_3COOC_2H_5]$ = 2/3 mole per liter; and $[H_2O]$ = 2/3 mole per liter. What is K_c?

$$K_c = \frac{[CH_3COOC_2H_5] \quad [H_2O]}{[C_2H_5OH] \quad [CH_3COOH]} = \frac{(2/3) \times (2/3)}{(1/3) \times (1/3)} = 4$$

Example No. 19-2. Calculate K_p for the reaction $N_2 + 3 H_2 \rightleftharpoons 2 NH_3$ (all gases) where the mixture at equilibrium at a temperature of 400°C and 1 atmosphere pressure contains 22.45% N_2, 67.35% H_2 and 10.20% NH_3 by volume.

First, we must calculate the partial pressure of each gas present in the mixture. According to Dalton's Law of partial pressures, each gas exerts a pressure proportional to its concentration,

partial pressure of N_2 = 22.45% of 1 atm = 0.2245 atmospheres

partial pressure of H_2 = 67.35% of 1 atm = 0.6735 atmospheres,

partial pressure of NH_3 = 10.20% of 1 atm = 0.1020 atmospheres.

_____

Total pressure = 1.0000 atmospheres

For simplicity, we will omit the units in the calculation, so

Then, $K_p = \dfrac{P^2_{NH_3}}{P_{N_2} \cdot P^3_{H_2}} = \dfrac{(0.1020)^2}{(0.2245)(0.6735)^3} = 0.1510$

Example No. 19-3. When it has reached equilibrium at a certain temperature and 3 atmospheres pressure, the reaction

$$H_2 + I_2 \rightleftharpoons 2 HI \text{ (all gases)}$$

contains 2.0 mol/L H_2, 3.0 mol/L I_2, and 17 mol/L HI. Calculate K_p.

Before we can calculate K_p we need to know the partial pressure of each gas. Assume a volume of 1 L.

$$\text{partial pressure} = \frac{\text{number of moles of the gas}}{\text{total number of moles present}} \times \text{total pressure,}$$

so partial pressure H_2, P_{H_2}, $= \dfrac{2.0 \text{ moles}}{2.0 + 3.0 + 17 \text{ moles}} \times 3.0 \text{ atm} = \dfrac{2.0}{22} \times 3.0 \text{ atm.}$

partial pressure I_2, p_{I_2}, $= \dfrac{3.0 \text{ moles}}{2.0 + 3.0 + 17 \text{ moles}} \times 3.0 \text{ atm} = \dfrac{3.0}{22} \times 3.0 \text{ atm}.$

partial pressure HI, P_{HI}, $= \dfrac{17 \text{ moles}}{2.0 + 3.0 + 17 \text{ moles}} \times 3.0 \text{ atm} = \dfrac{17}{22} \times 3.0 \text{ atm}.$

Again, omitting units,

Then, $K_p = \dfrac{P_{HI}{}^2}{P_{H_2} \, P_{I_2}} = \dfrac{\left(\dfrac{17}{22} \times 3.0\right)^2}{\left(\dfrac{2.0}{22} \times 3.0\right)\left(\dfrac{3.0}{22} \times 3.0\right)} = 48.$

Example No. 19-4. 1.00 mole PCl_5 was placed in a 4.00 liter container and heated to 230°C. Calculate the concentrations of PCl_3, Cl_2, and PCl_5 at equilibrium if $K_c = 49.0$.

$$PCl_3 + Cl_2 \rightleftharpoons PCl_5$$

Let us assume that X moles per liter of PCl_5 dissociated. Then there will be X moles per liter of PCl_3 formed and also X moles per liter of Cl_2 at equilibrium since the equation tells us that one mole of PCl_3 and one mole Cl_2 are formed for every mole PCl_5 that dissociates. The original PCl concentration was 0.250 moles/liter (1.00 mole in 4.00 liters) so that the concentration of the PCl_5 remaining will be 0.250 - X moles/liter.

Then, setting up the equilibrium constant, and eliminating units for simplicity.

$$\dfrac{[PCl_5]}{[PCl_3] \quad [Cl_2]} = K_c \qquad\qquad \dfrac{[0.250-X]}{[X] \quad [X]} = 49.0$$

Cross multiplying, $0.250 - X = 49 X^2$ and so

$$49 X^2 + X - 0.250 = 0$$

Solving this equation by the quadratic formula, $X = \dfrac{-b \pm \sqrt{b^2 - 4ac}}{2a}$

$$X = \dfrac{-1 \pm \sqrt{1^2 - 4 \times 49.0 \times (-0.250)}}{2 \times 49.0} = \dfrac{-1 \pm \sqrt{50.0}}{98.0}$$

$$X = \dfrac{-1 + 7.07}{98.0} \text{ and } X = \dfrac{-1 - 7.07}{98.0}$$

$$X = 0.0619 \qquad\qquad X = -0.0823$$

We have two possible answers for this quadratic equation. However, only one of these answers is possible because we cannot have a negative concentration. Therefore,

$[PCl_3] = 0.0619$ moles/liter, $[Cl_2] = 0.0619$ moles/liter,

$[PCl_5] = 0.250 - 0.0619 = 0.188$ moles/liter

Other examples of equilibrium reactions are discussed in Chapters 21 and 22, ionization reactions and solubility product constant.

Relationship between K_p and K_c

K_p and K_c are interrelated as indicated by the formula

$$K_p = K_c \, (RT)^{\Delta n}$$

where Δn is the change in the number of moles of gaseous products in an equilibrium reaction. Δn equals the number of moles of gaseous products minus the number of moles of gaseous reactants. In the reaction

$$2\,O_{3(g)} \rightleftarrows 3O_{2(g)}$$

$$n = 3 - 2 = 1$$

If the units of gas concentration are mol/L and temperature is in Kelvins, the value of R, the universal gas constant (see page 38) is 0.0821 L·atm/mol·K

Example 19-5. For the reaction $2\,NH_{3(g)} \rightleftarrows N_{2(g)} + 3H_{2(g)}$ $K_c = 30.0$ at 500°C

Using the formula $K_p = K_c\,(RT)^{\Delta n}$ where $K_c = 30.0$, $R = 0.0821$, $T = 773K$, and $\Delta n = 2$

$K_p = 30.0\,(0.0821 \times 773)^2 = 1.21 \times 10^5$

Example 19-6. K_c for the reaction $PCl_{3(g)} + Cl_{2(g)} \rightleftarrows PCl_{5(g)}$

is 49.0 at 230°C. What is the value of K_p?

$K_p = K_c\,(RT)^{\Delta n}$ where $\Delta n = -1\,(1 - 2)$,

$K_p = 49.0\,(0.0821 \times 503)^{-1} = 1.19$

Problem Assignment

What will be the effect upon the equilibrium and upon the equilibrium constant of the following reactions at equilibrium if:

a. the temperature is increased
b. the pressure is decreased
c. a catalyst is added
d. the concentration of the underlined substance is increased.

19-1. $NO + 1/2\,\underline{O_2} \rightleftarrows NO_2$ $\Delta H = -56.4$ kJ (all gases)

19-2. $CO_2 + H_2 \rightleftarrows \underline{CO} + H_2O$ $\Delta H = 41.8$ kJ (all gases)

19-3. $\underline{H_2} + 1/2\,O_2 \rightleftarrows H_2O$ $\Delta H = -242$ kJ (all gases)

19-4. $\underline{Cl_2}_{(gas)} + 2\,KI\,_{(aqueous)} \rightleftarrows 2\,KCl\,_{(aqueous)} + I_{2(solid)}$ $\Delta H = -109$ kJ

19-5. $1/2\,N_2 + 3/2\,H_2 \rightleftarrows \underline{NH_3}$ $\Delta H = -46$ kJ (all gases)

19-6. $SO_3 \rightleftarrows \underline{SO_2} + 1/2\,O_2$ $\Delta H = 94$ kJ (all gases)

19-7. $2\,H_2O + 2\,\underline{Cl_2} \rightleftarrows 4\,HCl + O_2$ $\Delta H = 30.1$ kJ/mole HCl (all gases)

19-8. $1/2\,\underline{N_2} + 1/2\,O_2 \rightleftarrows NO$ $\Delta H = 180$ kJ (all gases)

19-9. $2\,O_3 \rightleftarrows 3\,\underline{O_2}$ $\Delta H = -132$ kJ/mole (all gases)

For simplicity, omit the units of K_c and K_p for the following problems:

19-10. Give the formula for K_c and K_p for Problems 19-5 through 19-9.

19-11. Calculate K_c for the reaction $A + 3B \rightleftarrows 2C + 3D$ if a 5.00 liter container of the mixture at equilibrium contains A in the concentration of 3.00 moles/liter; B, 2.00 moles/liter; C, 4.00 moles/liter; and D, 1.00 mole/liter. . . . . (0.667)

19-12. Calculate K_c for the reaction $2\,N_2 + 5\,O_2 \rightleftarrows 2\,N_2O_5$ (all gases) if a 20.0 liter container holds 6.00 moles of N_2, 2.00 moles O_2, and 3.00 moles N_2O_5. . . . . (2.50×10^4)

19-13. If the N_2 concentration in Problem 19-12 becomes 0.0500 moles/liter, and the O_2 0.100 moles/liter, what will be the concentration of the N_2O_5? . . . . (2.50×10^{-2} moles/liter)

19-14.   Calculate K_c and K_p for the reaction $2\,SO_2 + O_2 \rightleftarrows 2\,SO_3$ if a 10.0 liter container at 0.750 atmospheres pressure holds 4.00 moles SO_2, 13.0 moles O_2 and 5.00 moles SO_3. ..... (1.20, 3.53)

19-15.   Calculate K_c for the reaction $PCl_5 \rightleftarrows PCl_3 + Cl_2$ (all gases) where a 20.0 liter container at equilibrium holds 10.0 moles PCl_5, 4.00 moles PCl_3, and 2.00 moles Cl_2.

19-16.   Calculate K_p for Problem 19-15 if the total pressure is 4.00 atm.

19-17.   A 5.00 liter container holds 1.00 mole N_2, 2.00 moles H_2, and 10.0 moles NH_3. Calculate K_c.
$N_2 + 3\,H_2 \rightleftarrows 2\,NH_3$.

19-18.   A container at a temperature of 500°C holds 5.20 moles NH_3, 44.8 moles N_2, and 38.6 moles H_2 at a pressure of 100 atm. Calculate K_p.
$2\,NH_3 \rightleftarrows N_2 + 3\,H_2$.

19-19.   Calculate K_c for the reaction $N_2 + O_2 \rightleftarrows 2\,NO$ if a 2.00 liter container holds 0.750 moles N_2, 0.750 moles O_2, and 0.0500 moles NO.

19-20.   If the number of moles of N_2 in Problem 19-19 becomes 0.500 and the number of moles of O_2 0.800, what will be the number of moles of NO present, assuming constant temperature?

19-21.   4.00 moles HI are placed in a 5.00 liter container and allowed to reach equilibrium.
$2\,HI \rightleftarrows H_2 + I_2$. How many moles of H_2, I_2 and HI will be present?
$K_c = 2.04 \times 10^{-2}$

19-22.   2.0 moles N_2O_4 are placed in a 4.0 liter container at 8°C. What will be the concentration of the N_2O_4 and the NO_2 at equilibrium if $K_c = 9.52 \times 10^4$.
$2\,NO_2 \rightleftarrows N_2O_4$.

19-23.   3.00 moles N_2 are mixed with 4.00 moles O_2 and allowed to reach equilibrium in a 1.00 liter container. How many moles per liter of NO are formed? How many moles per liter of N_2 are O_2 are left?
$2\,NO \rightleftarrows N_2 + O_2 \qquad K_c = 2.22 \times 10^2$.

19-24.   The equilibrium constant for the reaction ethyl alcohol + acetic acid $\rightleftarrows$ ethyl acetate + water is 4.00. What will be the concentrations of the ethyl alcohol and the acetic acid at equilibrium when the alcohol and the acid are mixed, each in the concentration of 2.00 moles per liter?

19-25.   A 4.0 liter container holds 3.2 moles PCl_3, 0.20 moles Cl_2, and 4.0 moles PCl_5 at equilibrium. If the number of moles of PCl_3 is increased to 4.0, what will be the concentrations at equilibrium?
$PCl_5 \rightleftarrows PCl_3 + Cl_2$.

19-26.   6.00 moles CO_2 are placed in a 5.00 liter container and heated to a temperature at which it was found to be 15.0% dissociated. Calculate K_c.
$2\,CO_2 \rightleftarrows 2\,CO + O_2$.

19-27.   An equilibrium mixture holds 1.50 moles H_2, 8.00 moles CO_2, 1.00 mole H_2O, and 3.00 moles CO in a 5.00 liter container. How much H_2 must be added to the container in order to increase the amount of CO to 3.50 moles?
$CO_2 + H_2 \rightleftarrows CO + H_2O$.

19-28.   4.00 moles H_2O are placed in a 2.00 liter container and heated to a temperature at which the H_2O is 1.00% dissociated. Calculate K_c.

19-29. K_c for the reaction $A_{(g)} + 3B_{(g)} \rightleftarrows 2C_{(g)} + 3D_{(g)}$ is 0.667 at 400 K. What is the value of K_p?

19-30. K_p for the gaseous reaction $X_{(g)} + Y_{(g)} \rightleftarrows 4Z_{(g)}$ is 6.53. K_c for the same reaction is 2.69×10^{-3}. What is the temperature in °C?

Chapter 20

HYDROGEN ION CONCENTRATION –pH

In expressing the concentration of the hydrogen ion and the hydroxide ion in solution we use "p," the mathematical symbol for the logarithm of the reciprocal or for the negative logarithm. Thus, pH is defined as the logarithm of the reciprocal of the hydrogen ion concentration or as the negative logarithm of the hydrogen ion concentration, all concentrations being expressed in moles per liter.

Thus, $pH = \log \dfrac{1}{[H^+]} = -\log [H^+]$ since

$\log \dfrac{1}{[H^+]} = \log 1 - \log [H^+]$ and the log 1 = 0.

The bracket [ ] indicates concentration expressed in moles per liter.

If the hydrogen ion concentration = 10^{-5} moles per liter,

$$pH = \log \frac{1}{10^{-5}} = \log \frac{10^5}{1} = 5$$

Similarly, pOH is the $\log \dfrac{1}{[OH^-]}$. If the hydroxide ion concentration = 10^{-10} moles per liter,

$$pOH = \log \frac{1}{10^{-10}} = \log \frac{10^{10}}{1} = 10.$$

We know that $[H^+][OH^-] = 10^{-14}$ at 25°C so that **pH + pOH = 14.**

Thus, we see that a low pH indicates an acid solution (high hydrogen ion concentration) while a high pH indicates a basic solution (low hydrogen ion concentration).

H^+ concentration in moles/liter		pH	
10^0	or 1	0	strongly acidic
10^{-1}	or 0.1	1	strongly acidic
10^{-2}	or 0.01	2	strongly acidic
10^{-3}	or 0.001	3	weakly acidic
10^{-4}	or 0.0001	4	weakly acidic
10^{-5}	or 0.00001	5	weakly acidic
10^{-6}	or 0.000001	6	weakly acidic
10^{-7}	or 0.0000001	7	neutral
10^{-8}	or 0.00000001	8	weakly basic
10^{-9}	or 0.000000001	9	weakly basic
10^{-10}	or 0.0000000001	10	weakly basic
10^{-11}	or 0.00000000001	11	weakly basic
10^{-12}	or 0.000000000001	12	strongly basic
10^{-13}	or 0.0000000000001	13	strongly basic
10^{-14}	or 0.00000000000001	14	strongly basic

Example No. 20-1.   If the hydrogen ion concentration is 0.000001 moles per liter, what is the pH? the pOH?

$$pH = \log \frac{1}{[H^+]} = \log \frac{1}{0.000001} = \log \frac{1}{10^{-6}} = \log \frac{10^6}{1} = 6$$

Since pH + pOH = 14, pOH = 14 – pH = 14 – 6 = 8.

Example No. 20-2.   What is the pH of a solution whose $[H^+] = 6.5 \times 10^{-10}$?

$$pH = \log \frac{1}{6.5 \times 10^{-10}} = \log \frac{10^{10}}{5.5} = \log 10^{10} - \log 6.5 = 10.00 - 0.81 = \underline{9.19}$$

($\log 6.5 = 0.81$ as obtained from a scientific calculator.

Example No. 20-3.   Calculate the pH of a 0.10 molar solution of acetic acid which is 1.3 percent ionized.

$[H^+] = 1.3\%$ of 0.10 moles/liter since it is only 1.3% ionized.

Therefore, $[H^+] = 0.013 \times 0.10$ moles/liter $= 1.3 \times 10^{-3} \dfrac{\text{moles}}{\text{liter}}$.

$$pH = \log \frac{1}{1.3 \times 10^{-3}} = \log \frac{10^3}{1.3} = \log 10^3 - \log 1.3 = 3.00 - 0.11 = \underline{2.89}.$$

Example No. 20-4.   A solution has a pH of 7.89. What is the hydrogen ion concentration expressed in moles per liter?

$$pH = \log \frac{1}{[H^+]} = 7.89$$

$\log 1 - \log [H^+] = 7.89$ but $\log 1 = 0$, so

$- \log[H^+] = 7.89$

$\log [H^+] = -7.89 = -8.00 + 0.11$, and

$[H^+] = $ antilog $(-8.00 + 0.11) = $ antilog $(-8.00) \times$ antilog (0.11)

antilog $(-8.00) = 10^{-8}$ and antilog $(0.11) = 1.3$, so

$[H^+] = 1.3 \times 10^{-8}$ moles per liter.

Example No. 20-5.   What is the percent ionization of a 0.10 molar acid HX solution whose pH is 2.5?

First, let us find the hydrogen ion concentration in moles/liter.

$$pH = \log \frac{1}{[H^+]}$$

$$2.5 = \log \frac{1}{[H^+]} = \log 1 - \log [H^+] = - \log[H^+] \quad \text{since } (\log 1 = 0)$$

so $\log[H^+] = -2.5 = -3.0 + 0.5$ and

$[H^+] = $ antilog $(-3.0 + 0.5) = $ antilog $(-3.0) \times$ antilog $(0.5) = 3.2 \times 10^{-3}$

The acid solution is 0.10 molar or there would be 0.10 moles per liter of H^+ ion if it were completely ionized. Therefore,

the percent ionization $= \dfrac{H^+ \text{ ion concentration}}{\text{original concentration}} \times 100$ so

percent ionization $= \dfrac{3.2 \times 10^{-3}}{0.10} \times 100 = 3.2\%$

Problem Assignment

20-1.   Calculate the pH and the pOH of a solution whose hydrogen ion concentration in moles per liter is:

a.	0.00001	f.	0.00045
b.	0.001	g.	3.58×10^{-8}
c.	10^{-11}	h.	1.15×10^{-2}
d.	1	i.	8.26×10^{-7}
e.	0.0000000001	j.	4.70×10^{-13}

> a. 5, 9
>
> f. 3.35, 10.65

20-2.   What is the hydrogen concentration of a solution whose pH is:

a.	2	f.	1.69
b.	5	g.	8.04
c.	14	h.	12.76
d.	2	i.	4.39
e.	0	j.	0.67

Are these solutions acid, basic, or neutral?

> a. 10^{-2}M, acid
>
> f. 2.04×10^{-2}M, acid

20-3.   Calculate the molarity of the following solutions from their respective pH's:

a.	HNO_3, pH 3.0	d.	H_2SO_4, pH 3.50
b.	KOH, pH 12.0	e.	$Ca(OH)_2$, pH 9.67
c.	$Ba(OH)_2$, pH 9.0	f.	HCl, pH 1.26

> d. $1.58 \times 10^{-4} M$

20-4.   Calculate the pH of a 0.0010 molar solution of an acid HX which is 30% ionized. . . . . (3.52)

20-5.   Calculate the pH of a 0.040 M solution of a base ZOH which is 1.5% ionized.

20-6.   The pH of a 1.00×10^{-2} M solution of an acid HX is 4.60. What is its percent ionization?   (0.251%)

20-7.   What is the pH of a 0.50 M solution of: a) NaOH, b) $Ca(OH)_2$, c) HNO_3, d) H_2SO_4?

20-8.   A solution containing 200 mg H_2SO_4 per 50 mL will have what pH?

20-9.   What is the pH of a solution containing 2.50 g NaOH per 400 mL of solution?

20-10.   What is the pH of a solution prepared by diluting 30 mL of concentrated sulfuric acid, density 1.86 g/mL, 98% by weight H_2SO_4, to 150 mL?

20-11.   Complete the following table:

	M	H^+	OH^-	pH	pOH
HCl	0.10				
KOH				12	
H_2SO_4	0.0050				
$Ca(OH)_2$			3.0×10^{-5}		
HNO_3					10
H_2SO_4		3.0×10^{-6}			
$Ba(OH)_2$				8.39	

Chapter 21

IONIZATION REACTIONS

The ionization of a weak electrolyte is in reality an equilibrium reaction and as such may be treated according to the rules of equilibria. Thus, for a solution of ammonium hydroxide, the equilibrium reaction is

$$NH_4OH \rightleftharpoons NH_4^+ + OH^-$$

and the equilibrium constant is

$$K_i = \frac{(NH_4^+)\ (OH^-)}{(NH_4OH)} \text{ where } K_i \text{ is the ionization constant.}$$

For acetic acid,

$$HC_2H_3O_2 \rightleftharpoons H^+ + C_2H_3O_2^-$$

$$K_i = \frac{(H^+)\ (C_2H_3O_2^-)}{(HC_2H_3O_2)}$$

If we had written the ionization of acetic acid as a reaction of water with the acid molecule,

$$HC_2H_3O_2 + H_2O \rightleftharpoons H_3O^+ + C_2H_3O_2^-$$

then

$$K_i = \frac{(H_3O^+)\ (C_2H_3O_2^-)}{(HC_2H_3O_2)\ (H_2O)}$$

However, in all dilute solutions, the concentration of water may be considered a constant so that

$$K_i(H_2O) = \frac{(H_3O^+)\ (C_2H_3O_2^-)}{(HC_2H_3O_2)} = K_2$$

where K_2 is a product of the two constants K_i and (H_2O).

Also H^+ and H_3O^+ represent the same ion, a hydrated hydrogen ion, so that this equilibrium constant K_2 is identical with the ionization constant K_i previously determined. Thus, for the ionization of weak electrolytes in dilute solution we may omit the reaction with water in the equilibrium reaction and also in the ionization constant.

As with the equilibrium constant, the ionization constant is affected by temperature only.

COMMON ION EFFECT

If, to a solution of acetic acid, we add some solid sodium acetate, we will increase the acetate ion concentration and so shift the equilibrium to the left.

$$HC_2H_3O_2 \rightleftharpoons H^+ + C_2H_3O_2^-$$

thus decreasing the H^+ concentration. This is called the common ion effect whereby one ion common to both the solution and the added substance decreases the concentration of the other ion in solution. Thus, the addition of the sodium acetate will decrease the H^+ concentration and make the solution less acid (higher) pH.

Example No. 21-1. Calculate the ionization constant for HCN if a 0.010 *M* solution contains 2.19×10^{-6} mole H^+ per liter. $HCN \rightleftharpoons H^+ + CN^-$

For every mole H^+ that is produced, one mole CN^- must also be produced, so,

$$(H^+) = 2.19 \times 10^{-6} \text{ moles per liter}$$

$$(CN^-) = 2.19 \times 10^{-6} \text{ moles per liter}$$

For every mole of H^+ or CN^- that is produced, 1 mole of HCN must have ionized, so, if 2.19×10^{-6} moles HCN ionized and there were 0.010 moles to begin with, the remaining HCN concentration =

$$(HCN) = 0.010 - 2.19 \times 10^{-6}$$

Then

$$K_i = \frac{(H^+) \; (CN^-)}{(HCN)} = \frac{(2.19 \times 10^{-6}) \; (2.19 \times 10^{-6})}{(0.010 - 2.19 \times 10^{-6})}$$

$$= \frac{(2.19 \times 10^{-6})^2}{(0.010 - 0.00000219)} = \frac{(2.19 \times 10^{-6})^2}{(0.0099978)} = 4.8 \times 10^{-10}$$

Example No. 21-2. What is the hydroxide ion concentration in $0.0100 \; M \; NH_4OH$ if its ionization constant is 1.80×10^{-5}?

$$NH_4OH \rightleftharpoons NH_4^+ + OH^-$$

Since we want the OH^- concentration, we let X = the OH^- concentration = (OH^-),

then X = the NH_4^+ concentration = (NH_4^+)

because, according to the equilibrium reaction above, one NH_4^+ and one OH^- ion are produced for every molecule of NH_4OH that ionizes.

The NH_4OH concentration was 0.0100 molar and since X moles of NH_4^+ and X moles OH^- are produced, X moles per liter of it must have ionized leaving

the NH_4OH concentration = $0.0100 - X$

Then $K_i = \dfrac{(NH_4^+) \; (OH^-)}{(NH_4OH)}$ so

$$1.80 \times 10^{-5} = \frac{(X) \; (X)}{(0.0100 - X)} = \frac{X^2}{0.0100 - X}$$

and $X^2 + 1.80 \times 10^{-5} X - 1.80 \times 10^{-7} = 0$

Using the quadratic formula $\qquad X = \dfrac{-b \pm \sqrt{b^2 - 4ac}}{2a}$

$$X = \frac{-1.80 \times 10^{-5} \pm \sqrt{(1.80 \times 10^{-5})^2 - 4(1)(-1.80 \times 10^{-7})}}{2(1)}$$

$$= \frac{-1.80 \times 10^{-5} \pm \sqrt{3.24 \times 10^{-10} + 7.20 \times 10^{-7}}}{2}$$

$$= \frac{-0.0000180 \pm \sqrt{0.000000000324 + 0.000000720}}{2}$$

$$= \frac{-0.0000180 \pm \sqrt{0.000000720324}}{2}$$

$$= \frac{-0.0000180 \pm 0.000848}{2}$$

$$= \frac{0.000830}{2} \text{ or } \frac{-0.000866}{2}$$

$$= 0.000415 \text{ or } -0.000433$$

and since there can never be a negative concentration, the correct result is

$$X = 0.000415 = 4.15 \times 10^{-4} \quad \text{moles per liter.}$$

If we had neglected the value X in comparison with the relatively large value of 0.0100, we would have had

$$\frac{(X)\ (X)}{(0.0100)} = 1.80 \times 10^{-5}$$

whence

$$X^2 = (1.80 \times 10^{-5})(0.0100) = 1.80 \times 10^{-7}$$

so

$$X = 4.24 \times 10^{-4} \text{ moles per liter}$$

which is not significantly different from the result derived before. However, we can neglect the value of X only when it is small in comparison with some other relatively larger number.

Example No. 21-3   Calculate the OH⁻ concentration in 1.0 M NH₄OH. $K_i = 1.8 \times 10^{-5}$.

This is the same problem as the previous one except that the concentration of the NH₄OH is greater. So, we let

$$X = (OH^-)$$

$$X = (NH_4^+), \qquad \text{and}$$

$$1.0 - X = (NH_4OH)$$

$$K_i = \frac{(NH_4^+)\ (OH^-)}{(NH_4OH)} = \frac{(X)\ (X)}{(1.0-X)} = 1.8 \times 10^{-5}$$

and neglecting the value of X in comparison with such a large value of 1.0, we get

$$\frac{(X)\ (X)}{(1.0)} = 1.8 \times 10^{-5}$$

$$X^2 = 1.8 \times 10^{-5}$$

$$X = 4.3 \times 10^{-3} \text{ moles per liter} = (OH^-)$$

which we can see is relatively unimportant in comparison with an NH₄OH concentration of 1.0 M so that neglecting it was permissible.

Example No. 21-4.   0.050 molar acetic acid is 1.88% ionized. Calculate its ionization constant.

$$HC_2H_3O_2 \rightleftharpoons H^+ + C_2H_3O_2^-$$

Since the solution is 1.88% ionized, 1.88% of 0.050 moles per liter must have ionized = 0.0188 × 0.050 = 0.00094 = 9.4 × 10⁻⁴ moles per liter ionized.

For each mole HC₂H₃O₂ that is ionized, 1 mole H⁺ and 1 mole C₂H₃O₂⁻ are produced, so 9.4 × 10⁻⁴ moles acetic acid ionized will yield

$$9.4 \times 10^{-4} \text{ moles } H^+ \text{ per liter} = (H^+) \text{ and}$$

$$9.4 \times 10^{-4} \text{ moles } C_2H_3O_2^- \text{ per liter} = (C_2H_3O_2^-).$$

There were 0.050 moles HC₂H₃O₂ per liter originally and 9.4×10⁻⁴ moles of it per liter ionized, so the HC₂H₃O₂ concentration =

$$0.050 - 9.4 \times 10^{-4} \text{ moles per liter} = (HC_2H_3O_2)$$

$$K_i = \frac{(H^+)\ (C_2H_3O_2^-)}{(HC_2H_3O_2)} = \frac{(9.4 \times 10^{-4})\ (9.4 \times 10^{-4})}{(0.050 - 9.4 \times 10^{-4})} = \frac{(9.4 \times 10^{-4})^2}{(0.049)} = 1.8 \times 10^{-5}$$

Example No. 21-5. Calculate the percent ionization of 0.00100 M NH$_4$OH if its K_i = 1.80×10^{-5}.

$$NH_4OH \rightleftharpoons NH_4^+ + OH^-$$

If we let

$$X = (NH_4^+), \text{ then, also}$$

$$X = (OH^-) \text{ and}$$

$$0.00100 - X = (NH_4OH)$$

since, for every mole of NH$_4^+$ and OH$^-$ produced, 1 mole NH$_4$OH must have ionized.

Then
$$K_i = \frac{(NH_4^+)\ (OH^-)}{(NH_4OH)}$$

$$1.80 \times 10^{-5} = \frac{(X)\ (X)}{(0.00100 - X)} = \frac{X^2}{(0.00100 - X)}$$

We cannot neglect the X in comparison with 0.00100 because if we do, $\dfrac{X^2}{0.00100}$ = 1.80 × 10^{-5} and X^2 = 1.80×10^{-8} so X = 1.34×10^{-4} which is fairly close to 0.00100. Thus to solve the equation we must use the quadratic formula. So,

$$X^2 + (1.80\times10^{-5})X - 1.80\times10^{-8} = 0$$

$$X = \frac{-b \pm \sqrt{b^2 - 4ac}}{2a}$$

$$X = \frac{-1.80\times10^{-5} \pm \sqrt{(1.80\times10^{-5})^2 - (4)(1.80\times10^{-8})}}{2(1)}$$

$$= \frac{-1.80\times10^{-5} \pm \sqrt{(1.80\times10^{-5})^2 + 7.20\times10^{-8}}}{2}$$

$$= \frac{-0.0000180 \pm \sqrt{3.24\times10^{-10} + 7.20\times10^{-8}}}{2}$$

$$= \frac{-0.0000180 \pm \sqrt{0.0000000720}}{2}$$

$$= \frac{-0.0000180 \pm 0.000269}{2}$$

$$= \frac{0.000251}{2} \text{ or } \frac{-0.000287}{2}$$

and since there can never be a negative concentration,

$$X = \frac{0.000251}{2} = 0.000126 \text{ moles per liter} = [NH_4^+] = [OH^-]$$

and 0.000126 = moles per liter NH$_4$OH ionized

The percent ionization = $\dfrac{\text{number of moles ionized}}{\text{total number of moles present}} \times 100 = \dfrac{0.000126}{0.00100} \times 100 = 12.6\%$

Example No. 21-6.   0.10 mole NaCN is added to a liter of 0.010 M HCN solution. What effect will this have on the H^+ ion concentration?

$$K_i = 4.8 \times 10^{-10}$$

$$HCN \rightleftharpoons H^+ + CN^-$$

First, to find the H^+ ion concentration in the original solution, we

let $X = (H^+)$ and so, also (CN^-)

then $0.010 - X = (HCN)$

$$K_i = \frac{(H^+) \ (CN^-)}{(HCN)}$$

$$4.8 \times 10^{-10} = \frac{(X) \ (X)}{(0.010 - X)} = \frac{X^2}{(0.010 - X)}$$

and since X will be small in comparison with 0.010 (if we drop it, $\frac{X^2}{0.010} = 4.8 \times 10^{-10}$ so X is about 10^{-6} which is small in comparison with 0.010) we have

$$4.8 \times 10^{-10} = \frac{X^2}{0.010}$$

$$X^2 = (4.8 \times 10^{-10}) \ (0.010) = 4.8 \times 10^{-12}$$

$$X = 2.2 \times 10^{-6} \text{ moles per liter} = (H^+)$$

To find the concentration of H^+ after the addition of 0.10 mole NaCN, first we assume that the NaCN is completely ionized so that the CN^- concentration from the NaCN = 0.10 moles per liter.

Then, we let

X = new H^+ concentration = (H^+)

and X will also = (CN^-) (the CN^- concentration) but we have added 0.10 moles per liter CN^- so that the total CN^- concentration will be

$$(CN^-) = X + 0.10$$

The HCN was 0.010 moles per liter and since we have X moles per liter of H^+, X moles per liter of it must have ionized so that the amount HCN left =

$$(HCN) = 0.01 - X$$

Then $$K_i = \frac{(H^+) \ (CN^-)}{(HCN)}$$

$$4.8 \times 10^{-10} = \frac{(X) \ (X + 0.10)}{(0.010 - X)}$$

The addition of the common ion, CN^-, will decrease the H^+ concentration. Originally the H^+ was 2.2×10^{-6} so that the new value, which will be smaller, will certainly be insignificant in comparison with such large values as 0.10 and 0.010 so that by "rounding off" we have

$$4.8 \times 10^{-10} = \frac{(X) \ (0.10)}{(0.010)}$$

$$X = \frac{(4.8 \times 10^{-10}) \ (0.010)}{(0.10)} = 4.8 \times 10^{-11} \text{ moles per liter} = (H^+)$$

So that the addition of 0.10 moles NaCN will change the H^+ from 2.2×10^{-6} moles per liter to 4.8×10^{-11} moles per liter.

Example No. 21-7. What is the pH of 0.100 M acetic acid solution which is 1.33% ionized?

$$K_i = 1.8 \times 10^{-5}$$

$$HC_2H_3O_2 \rightleftharpoons H^+ + C_2H_3O_2^-$$

To find pH we need to know the H^+ concentration, so that is the first thing we must determine.

The solution is 1.33% ionized, so that 1.33% of 0.100 moles = 1.33% $\times$ 0.100 = 0.0133 $\times$ 0.100 = 0.00133 moles acetic acid ionized.

For every mole of acetic acid that ionizes (according to the equilibrium reaction) 1 mole H^+ is produced so that

$$0.00133 \text{ moles } H^+ \text{ per liter are produced, or}$$

$$(H^+) = 0.00133$$

$$\text{The pH} = \log \frac{1}{(H^+)} = \log \frac{1}{0.00133}$$

$$= \log \frac{1}{1.33 \times 10^{-3}} = \log \frac{10^3}{1.33}$$

$$= \log 10^3 - \log 1.33$$

$$= 3.00 - 0.12 = \underline{2.88}$$

Example No. 21-8. The pH of a 0.050 M acid solution (HA) is 3.60. What is the percent ionization?

$$HA \rightleftharpoons H^+ + A^-$$

First, let us find the H^+ concentration

$$pH = \log \frac{1}{(H^+)} = 3.60$$

So $\quad\quad 3.60 = \log 1 - \log (H^+) = -\log(H^+)$ (since log 1 = 0)

and $\quad\quad \log (H^+) = -3.60 = -4.00 + 0.40$

$$(H^+) = \text{antilog} (-4.00) \times \text{antilog} (0.40)$$

$$= 10^{-4} \times 2.5$$

$$= 2.5 \times 10^{-4} = \text{moles } H^+ \text{ per liter}$$

This is also the moles per liter of HA that are ionized, since for every mole H^+ produced 1 mole HA must have ionized.

Then, since % ionization =

$$\frac{\text{number of moles ionized}}{\text{total number of moles}} \times 100, \text{ the solution is}$$

$$\frac{2.5 \times 10^{-4}}{0.050} \times 100 = 0.50 \text{ percent ionized}$$

Example No. 21-9. The ionization constant of benzoic acid is 6.3×10^{-5}. Calculate the pH of a 0.10 M solution of it.

$$HC_7H_5O_2 \rightleftharpoons H^+ + C_7H_5O_2^-$$

To find pH we need to know (H^+), so we let

$$X = (H^+), \text{ then}$$

$$X = (C_7H_5O_2^-) \text{ and}$$

$$0.10-X = (HC_7H_5O_2)$$

$$K_i = \frac{(H^+)\ (C_7H_5O_2^-)}{(HC_7H_5O_2)}$$

$$6.3\times10^{-5} = \frac{(X)\ (X)}{(0.10-X)}$$

Neglecting the value of X in comparison with 0.10 (since we can see the X will be of the order of 10^{-3})

$$6.3\times10^{-5} = \frac{(X)\ (X)}{0.10} \text{ and}$$

$$X^2 = (6.3\times10^{-5})\ (0.10) = 6.3\times10^{-6}$$

$$X = 2.5\times10^{-3} = (H^+)$$

$$pH = \log\frac{1}{(H^+)} = \log\frac{1}{2.5\times10^{-3}} = \log\frac{10^3}{2.5}$$

$$= \log 10^3 - \log 2.4$$

$$= 3.00 - 0.40 = \underline{2.60}$$

Example No. 21-10. In Example 21-6, what effect did the addition of the NaCN have upon the pH?

The original (H^+) was 2.2×10^{-6}, so

$$pH = \log\frac{1}{(H^+)} = \log\frac{1}{2.2\times10^{-6}} = \log\frac{10^6}{2.2}$$

$$= \log 10^6 - \log 2.2$$

$$= 6.00 - 0.34 = \underline{5.66}$$

The final (H^+) was 4.8×10^{-11}

$$pH = \log\frac{1}{(H^+)} = \log\frac{1}{4.8\times10^{-11}}$$

$$= \log\frac{10^{11}}{4.8}$$

$$= \log 10^{11} - \log 4.8$$

$$= 11.00 - 0.68 = \underline{10.32}$$

Example No. 21-11. The ionization constant of boric acid H_3BO_3 is 5.8×10^{-10}. Calculate the molar strength of a boric acid solution whose pH is 4.86.

$$H_3BO_3 \rightleftharpoons H^+ + H_2BO_3^-$$

Since we know the pH, we can find the (H^+)

$$pH = \log \frac{1}{(H^+)}$$

$$4.86 = \log \frac{1}{(H^+)} = \log 1 - \log (H^+) = -\log(H^+)$$

$$\log(H^+) = -4.86 = -5.00 + 0.14$$

$$(H^+) = \text{antilog}(-5.00) \times \text{antilog}(0.14)$$

$$(H^+) = 10^{-5} \times 1.4 = 1.4 \times 10^{-5}$$

Since we want the boric acid concentration,

we let X = the H_3BO_3 concentration

The $(H^+) = 1.4 \times 10^{-5}$. The $H_2BO_3^-$ will be the same, $1.4 \times 10^{-5} = (H_2BO_3^-)$ and $(H_3BO_3) = (X - 1.4 \times 10^{-5})$ because for every mole H^+ produced, 1 mole H_3BO_3 must have ionized.

$$K_i = \frac{(H^+)\ (H_2BO_3^-)}{(H_3BO_3)}$$

$$5.8 \times 10^{-10} = \frac{(1.4 \times 10^{-5})\ (1.4 \times 10^{-5})}{(X - 1.4 \times 10^{-5})}$$

and neglecting 1.4×10^{-5} in comparison with X,

$$5.8 \times 10^{-10} = \frac{(1.4 \times 10^{-5})\ (1.4 \times 10^{-5})}{X}$$

$$X = \frac{(1.4 \times 10^{-5})\ (1.4 \times 10^{-5})}{5.8 \times 10^{-10}}$$

$$= 0.34 \text{ moles per liter}$$

$$= 0.34 \text{ molar solution}$$

Problem Assignment

21-1. Calculate the H^+ ion concentration and the pH of a 0.120 M solution of formic acid. $HCHO_2 \rightleftharpoons H^+ + CHO_2^-$. $K_i = 1.8 \times 10^{-4}$ .... (4.6×10^{-3}, 2.34)

21-2. What is the percent ionization of 0.100 M HNO_2? of 1.00 M HNO_2? $K_i = 4.5 \times 10^{-4}$. .... (6.7%, 2.1%)

21-3. What is the ionization constant of 0.10 M HCN which is 0.0009% ionized? ..... (4.8×10^{-10})

21-4. What is the pH of 0.500 M acetic acid solution? If 0.750 moles of sodium acetate are added to a liter of the 0.500 M acid, what will be the pH? $K_i = 1.8 \times 10^{-5}$. .... (2.52, 4.57)

21-5. What is the percent ionization of 0.150 M acetic acid solution? of 0.0150 M acetic acid solution? $K_i = 1.8 \times 10^{-5}$.

21-6. What is the pH and the percent ionization of 0.025 M hypochlorous acid solution ($HClO$)? $K_i = 3.0 \times 10^{-8}$.

21-7. What is the concentration of an acid HX which is 5.00% ionized and whose K_i is 3.50×10^{-5}? . . . . (1.33×10^{-2} M)

21-8. What is the pH and the percent ionization of 0.100 M NH_4OH? $K_i = 1.8 \times 10^{-5}$.

21-9. Calculate the pH of 1.00 liter of 0.100 M $HC_2H_3O_2$ solution. What will be the pH when (a) 1.00 mole $NaC_2H_3O_2$ is added; (b) 10.0 g $NaC_2H_3O_2$ are added; (c) 1.00 liter of 0.0500 M NaOH is added? $K_i = 1.8 \times 10^{-5}$.

21-10. What is the pH of 0.0500 M NH_4OH? If 0.500 moles/liter of NH_4Cl are added, what will be the pH? $K_i = 1.8 \times 10^{-5}$.

21-11. Calculate the K_i for a weak base XOH if a 0.150 M solution has an OH^- concentration of 1.0×10^{-3} M.

21-12. What is the ionization constant for a 1.0 M solution of:
a. a weak acid, HX, whose pH is 2.00. . . . . (1.0×10^{-4})
b. a weak base, XOH, whose pH is 12.00.
c. a weak acid, HX, whose pH is 2.67.
d. a weak base, BOH, whose pH is 10.43.

21-13. Calculate the percent ionization for each solution in Problem No. 21-12.

21-14. What is the percent ionization and the pH of the following hydrofluoric acid solutions? $K_i = 6.8 \times 10^{-4}$.

$HF \rightleftharpoons H^+ + F^-$ a. 1.00 M b. 0.100 M c. 0.0100 M d. 0.00100 M

21-15. What is the pH of a solution containing: (a) 14.4 g $HC_2H_3O_2$ per 500 mL solution? (b) 14.4 g. $HC_2H_3O_2$ and 3.28 g $NaC_2H_3O_2$ per 500 mL solution? . . . . (2.53, 3.97)

21-16. The K_i for HCNO is 3.5×10^{-4}. How many moles of NaCNO must be added to 500 mL of 0.200 M HCNO solution to give a pH of 4.50? How many grams?

21-17. What will be the pH of a solution containing 150 g NH_3 per 750 mL solution? $K_i = 1.8 \times 10^{-5}$.

21-18. What will be the pH of a solution prepared by mixing 300 mL of 0.100 M NH_4OH and 300 mL of 0.100 M NH_4Cl?

21-19. What will be the pH of a solution prepared by mixing 40.0 mL of 0.100 M $HC_2H_3O_2$ with 20.0 mL of 0.100 M $NaC_2H_3O_2$?

21-20. How many grams of NH_4Cl must be added to 500 mL of 0.100 M NH_4OH solution to give a pH of 8.50?

21-21. Which solution will have a higher pH? (a) 500 mL of 0.100 M $HC_2H_3O_2$ to which has been added 4.10 g $NaC_2H_3O_2$ or (b) 200 mL of 0.200 M $HC_2H_3O_2$ to which has been added 2.95 g $NaC_2H_3O_2$?

21-22. What will be the H^+ concentration and the pH of a solution prepared by mixing 20.0 mL of 6.00 M $HC_2H_3O_2$ with 10.0 mL of 3.00 M NaOH?

21-23. What will be the sulfide ion concentration in a saturated H_2S solution (0.10 M) if the H^+ concentration is 0.100 M?
$$H_2S \rightleftharpoons 2\,H^+ + S^{2-} \qquad K_i = 1.1 \times 10^{-22}.$$

21-24. What will be the pH of the solution in Problem No. 21-23 in order to give a S^{2-} concentration of 1.00×10^{-15}?

Chapter 22

SOLUBILITY PRODUCT CONSTANT

If we place some solid CuS in water, a small amount of it will dissolve, setting up an equilibrium reaction

$$CuS \text{ (solid)} \rightleftharpoons Cu^{2+} + S^{2-} \text{ (in solution)}$$

whose ionization constant is

$$K_i = \frac{[Cu^{2+}] \; [S^{2-}]}{[CuS]}$$

However, as we discussed in the chapter on equilibrium reactions, the concentrations of solids are constants, so that

$$[Cu^{2+}] \; [S^{2-}] = K_i \times [CuS] = K_{SP}$$

where K_{SP}, the solubility product constant, is a product of two other constants, K_i and $[CuS]$. Thus, K_{SP} for $CuS = [Cu^{2+}] \; [S^{2-}]$.

Following the same rules mentioned in the preceding chapter for obtaining K_i for various compounds, we find that K_{SP} for any compound is equal to the product of the concentration of the ions (in moles per liter), each concentration being raised to the power indicated by its coefficient in the ionization reaction or by its subscript in the formula.

Thus,

$$K_{SP} \text{ for } Mg(OH)_2 \quad = [Mg^{2+}] \; [OH^-]^2$$

$$K_{SP} \text{ for } BaCrO_4 \quad = [Ba^{2+}] \; [CrO_4^{2-}]$$

$$K_{SP} \text{ for } Fe(OH)_3 \quad = [Fe^{3+}] \; [OH^-]^3$$

and in general

$$K_{SP} \text{ for } X_n \, Y_m \quad = [X]^n \; [Y]^m$$

This rule holds true for slightly soluble salts only.

USE OF THE SOLUBILITY PRODUCT CONSTANT

a. In precipitation calculations

If the product of the ion concentrations (raised to the appropriate power) exceeds the corresponding solubility product constant, a precipitate will form (removing some of the ion from solution) until the new product of the concentrations does not exceed the solubility product constant. If the product of the ion concentrations (raised to the proper power) does not exceed the solubility product constant, no precipitate will form.

Thus, if we take a saturated solution of $CaCrO_4$ and add some K_2CrO_4 solution to it, we are increasing the concentration of the CrO_4^{2-} ion in solution. A saturated solution of $CaCrO_4$ indicates that $[Ca^{2+}] \; [CrO_4^{2-}]$ just equals K_{SP} but does not exceed it. Now, upon adding K_2CrO_4, we are increasing the CrO_4^{2-} concentration so that $[Ca^{2+}] \; [CrO_4^{2-}]$ is greater than K_{SP} and so a precipitate of $CaCrO_4$ will form. This precipitate will continue to form until $[Ca^{2+}] \; [CrO_4^{2-}]$ again does not exceed K_{SP}. This means that the solution will contain less Ca^{2+} than it did previously since the addition of CrO_4^{2-} drives the equilibrium $CaCrO_4 \rightleftharpoons Ca^{2+} + CrO_4^{2-}$ to the left.

b. In dissolving precipitates

If a solution contains a precipitate, this indicates that the product of the concentrations (each raised to the appropriate power) must exceed the corresponding K_{SP}. To dissolve this precipitate, we must alter the concentrations so that K_{SP} is not exceeded. Thus, for

$$Zn(OH)_2 \rightleftharpoons Zn^{2+} + 2\ OH^-$$

$$K_{SP} = [Zn^{2+}]\ [OH^-]^2$$

If we reduce the concentration of the OH^- so that K_{SP} is not exceeded, the precipitate will begin to dissolve. Thus, if we add an acid (such as HCl, H_2SO_4, etc.) to supply H^+ ions, the H^+ will react with the OH^- to form H_2O, removing the OH^- from the solution. Removal of some of the OH^- ions from solution will cause the equilibrium point to shift to the right and will cause more $Zn(OH)_2$ to dissolve

$$Zn(OH)_2 \rightleftharpoons Zn^{2+} + 2\ OH^-$$

Example No. 22-1.   Calculate the solubility product constant for silver chloride if its solubility is 1.92×10^{-3} grams per liter at 25°C.

$$AgCl_{(solid)} \rightleftharpoons Ag^+ + Cl^-_{(in\ solution)}$$

To calculate K_{SP} we need the concentration of the ions in moles per liter, not in grams per liter. So since the molar mass of AgCl is 143 grams,

$$\frac{1.92\ x\ 10^{-3}\ grams\ per\ liter}{143\ grams\ per\ mol} = 1.34\ x\ 10^{-5}\ mol/L$$

According to the equilibrium reaction, each mole of AgCl that dissolves yields 1 mole Ag^+ and 1 mole Cl^-, so

1.34×10^{-5} moles AgCl that dissolve per liter will yield

1.34×10^{-5} moles Ag^+ per liter, and

1.34×10^{-5} moles Cl^- per liter.

Then, since $K_{SP} = [Ag^+]\ [Cl^-]$

$$K_{SP} = (1.34 \times 10^{-5})\ (1.34 \times 10^{-5}) = 1.80 \times 10^{-10}$$

Example No. 22-2.   Calculate the solubility product constant (at 25°C) for magnesium hydroxide if its solubility is 7.37×10^{-5} moles per liter.

$$Mg(OH)_{2(solid)} \rightleftharpoons Mg^{2+} + 2\ OH^-_{(in\ solution)}$$

One mole $Mg(OH)_2$, when dissolved, yields one mole of Mg^{2+} and two (2) moles OH^-, so 7.37×10^{-5} moles $Mg(OH)_2$ per liter will yield

7.37×10^{-5} moles per liter of Mg^{2+} and

$2 \times 7.37 \times 10^{-5} = 14.7 \times 10^{-5}$ moles per liter of OH^-.

Then, since $K_{SP} = [Mg^{2+}]\ [OH^-]^2$

$$K_{SP} = (7.37 \times 10^{-5})\ (14.7 \times 10^{-5})^2$$

$$K_{SP} = 1.60 \times 10^{-12} = 1.6 \times 10^{-12}\ (rounded\ off)$$

Example No. 22-3. The solubility product constant for ZnS at 20°C is 2×10^{-25}. Calculate the molar solubility of ZnS at 25°C.

$$ZnS_{(solid)} \rightleftharpoons Zn^{2+} + S^{2-}{}_{(in\ solution)}$$

Since we want the molar solubility of ZnS, we

let X = the molar solubility of ZnS.

Each mole of ZnS that dissolves yields 1 mole of Zn^{2+} and 1 mole S^{2-}, so X moles of ZnS that dissolve will yield

X moles Zn^{2+} per liter and

X moles S^{2-} per liter of solution.

Then, since $K_{SP} = [Zn^{2+}]\ (S^{2-}]$

$2 \times 10^{-25} = (X)\ (X)$ and

$2 \times 10^{-25} = X^2$ and

$X = 4.5 \times 10^{-13}$ moles per liter

Thus, the molar solubility of ZnS is 4.5×10^{-13}.

Example No. 22-4. Calculate the solubility in grams per liter of $PbCl_2$ if its solubility product constant at 25°C is 1.7×10^{-5}.

$$PbCl_{2(solid)} \rightleftharpoons Pb^{2+} + 2\ Cl^{-}{}_{(in\ solution)}$$

First, we **let X = the molar solubility of the $PbCl_2$** (since concentrations are easier to work with if they are in moles per liter).

Each mole of $PbCl_2$ that dissolves yields 1 mole Pb^{2+} and 2 moles Cl^-, so X moles $PbCl_2$ that dissolve in a liter will yield

X moles Pb^{2+} per liter and

$2X$ moles Cl^- per liter.

Since $K_{SP} = [Pb^{2+}]\ [Cl^-]^2$,

$1.7 \times 10^{-5} = (X)\ (2\ X)^2$

$1.7 \times 10^{-5} = 4\ X^3$

$0.42 \times 10^{-5} = X^3$

$\sqrt[3]{0.42 \times 10^{-5}} = X = 1.6 \times 10^{-2}$

so the molar solubility of $PbCl_2$ is 3.6×10^{-2}. However, the problem asks for the solubility in grams per liter, and since the molar mass of $PbCl_2$ is 278 grams,

$$1.6 \times 10^{-2} \frac{moles}{liter} \times 278 \frac{grams}{mole} = 4.4 \frac{grams}{liter} PbCl_2.$$

Example No. 22-5. The carbonate concentration of a solution is 3.00×10^{-2} moles per liter. Calculate the concentration of Ba^{2+} that must be exceeded in order to precipitate $BaCO_3$. K_{SP} for $BaCO_3 = 5.0 \times 10^{-9}$.

$$BaCO_{3(solid)} \rightleftharpoons Ba^{2+} + CO_3^{2-} \text{ (in solution)}$$

First, $$K_{SP} = [Ba^{2+}] [CO_3^{2-}]$$

$$K_{SP} = 5.0 \times 10^{-9}$$

$$[CO_3^{2-}] = 3.00 \times 10^{-2} \text{ moles per liter and}$$

$$[Ba^{2+}] = X \text{ moles per liter}$$

Then, $$5.0 \times 10^{-9} = (X) (3.00 \times 10^{-2}) \text{ whence}$$

$$X = \frac{5.0 \times 10^{-9}}{3.00 \times 10^{-2}} = 1.7 \times 10^{-7}$$

Thus, the concentration of the barium ions must exceed 1.7×10^{-7} moles per liter before a precipitate of $BaCO_3$ can form.

Example No. 22-6. To a liter of a solution containing a precipitate of silver chloride, 0.010 moles solid NaCl was added. How did the concentration of the silver ion in solution change? K_{SP} for $AgCl = 1.8 \times 10^{-10}$.

First, we must calculate the concentration of the silver ion in the original solution and second, in the new solution containing the NaCl.

$$AgCl \rightleftharpoons Ag^+ + Cl^-$$

and $K_{SP} = [Ag^+] [Cl^-] = 1.8 \times 10^{-10}$

For the original solution,

Let X = the concentration of Ag^+ ions in moles per liter, so
X = the concentration of the Cl^- ions in moles per liter.

$$(X) (X) = 1.8 \times 10^{-10}$$

$$X^2 = 1.8 \times 10^{-10}$$

$$X = 1.3 \times 10^{-5} \text{ moles per liter of } Ag^+ \text{ ion in the original solution.}$$

For the new solution,

Let X = the concentration of the Ag^+ ions in moles per liter. Then X will also equal the concentration of the Cl^- ion but we have added 0.010 moles NaCl per liter and since the NaCl is completely ionized, the solution will contain 0.010 moles Na^+ per liter and 0.010 moles Cl^- per liter. Thus, the total Cl^- concentration will be $(0.010 + X)$ moles per liter.

Then, $$[Ag^+] [Cl^-] = K_{SP} \text{ so } (X) (0.010 + X) = 1.8 \times 10^{-10}$$

This problem can be solved in two different ways. We can obtain a quadratic equation from the above calculation or we can do some "rounding off" of the concentrations.

We know that the Ag^+ ion concentration and also the Cl^- ion concentration in the original solution was 1.05×10^{-5} and now the Ag^+ concentration will certainly be less because the addition of more Cl^- ions will shift the equilibrium to the left,

$$AgCl \rightleftharpoons Ag^+ + Cl^-$$

so that X (the Ag^+ concentration) will be a very small number, certainly less than 1×10^{-5}. We can neglect this number in comparison with such a relatively large number such as 0.01. Thus we have

$$(X)(0.010) = 1.8 \times 10^{-10}$$

$$0.010\,X = 1.8 \times 10^{-10}$$

$X = 1.8 \times 10^{-8}$ moles per liter so that the addition of 0.010 moles solid NaCl per liter changed the Ag^+ concentration from 1.3×10^{-5} to 1.8×10^{-8} moles per liter.

If we had solved the quadratic equation

$$X(0.010 + X) = 1.8 \times 10^{-10}$$

$$X^2 + 0.010\,X - 1.8 \times 10^{-10} = 0 \text{ by using the quadratic formula}$$

$$X = \frac{-b \pm \sqrt{b^2 - 4ac}}{2a} \text{ we would have}$$

$$X = \frac{-0.010 \pm \sqrt{(0.010)^2 - 4\,(1)\,(-1.8 \times 10^{-10})}}{2\,(1)}$$

$$= \frac{-0.010 \pm \sqrt{0.00010 + 0.00000000072}}{2}$$

$$= \frac{-0.010 \pm \sqrt{0.00010000072}}{2}$$

$$= \frac{-0.010 \pm 0.010000036}{2}$$

$$= \frac{0.000000036}{2} \text{ or } \frac{-0.020000036}{2}$$

and since we can never have a negative concentration, we omit the second result so

$$X = \frac{0.000000036}{2} = 1.8 \times 10^{-8}$$

Thus the Ag^+ concentration in moles per liter is 1.8×10^{-8} which is identical with the result we obtained above.

Example No. 22-7. 1.0 liter of 0.10 molar NH_4OH solution and 1.0 liter of 0.10 molar $MgCl_2$ are mixed. Will a precipitate of $Mg(OH)_2$ form?

$$K_{SP} \text{ for } Mg(OH)_2 = 1.6 \times 10^{-12}$$

$$K_i \text{ for } NH_4OH = 1.8 \times 10^{-5}$$

First, since we are mixing the two solutions, the final volume will be 2.0 liters so that each concentration will be halved (0.10 mole NH_4OH per 2.0 liters of solution = 0.050 molar and 0.10 mole $MgCl_2$ per 2.0 liters of solution = 0.050 molar).

Since $NH_4OH \rightleftharpoons NH_4^+ + OH^-$

$$K_i = \frac{[NH_4^+]\ [OH^-]}{[NH_4OH]} = 1.8 \times 10^{-5}$$

Then, we let X = the concentration of the OH^- ions

so X = the concentration of the NH_4^+ ions

and $0.050 - X$ = the concentration of the NH_4OH

$$K_i = \frac{[NH_4^+]\ [OH^-]}{[NH_4OH]} = 1.8 \times 10^{-5} = \frac{(X)\ \ \ \ \ (X)}{(0.050 - X)}$$

and neglecting X in comparison with 0.050,

$$\frac{(X)\ (X)}{0.050} = 1.8 \times 10^{-5}$$

$$X^2 = (1.8 \times 10^{-5})\ (0.050) = 9.0 \times 10^{-7}$$

$$X = 9.5 \times 10^{-4} \text{ moles per liter} = [OH^-]$$

Since the $MgCl_2$ is completely ionized, $[Mg^{2+}] = 0.050$ moles per liter.

The product for $Mg(OH)_2 = [Mg^{2+}]\ [OH^-]^2 = (0.050)\ (9.5 \times 10^{-4})^2 = 4.5 \times 10^{-8}$ which exceeds the solubility product of $Mg(OH)_2$ (1.6×10^{-12}) so that *a precipitate will form.*

Example No. 22-8.   What concentration of NH_4Cl must be added to the solution in the previous example in order to prevent the precipitate from forming?

The addition of NH_4Cl will produce more NH_4^+ ions in solution and so drive the equilibrium reaction $NH_4OH \rightleftharpoons NH_4^+ + OH^-$ to the left thus reducing the OH^- concentration so that K_{SP} for $Mg(OH)_2$ will not be exceeded.

First, let us see what the OH^- concentration should be when the product of the ion concentrations equals the solubility product constant.

$$K_{SP} = 1.6 \times 10^{-12} = [Mg^{2+}]\ [OH^-]^2$$

and from the above problem $[Mg^{2+}] = 0.050$ moles per liter, so letting $X = [OH^-]$

1.6 $2.8 \times 10^{-12} = (0.050)\ (X)^2$

$$X^2 = \frac{1.6 \times 10^{-12}}{0.050} = 3.2 \times 10^{-11}$$

$$X = 5.7 \times 10^{-6} = OH^- \text{ concentration which must be exceeded in order for a}$$

precipitate to form.

Thus, in the 0.050 molar NH_4OH solution, the OH^- concentration $= 5.7 \times 10^{-6}$ which must also be the NH_4^+ concentration (due to the NH_4OH) since each mole of NH_4OH that ionizes produces 1 mole NH_4^+ and 1 mole OH^-. The NH_4OH concentration remaining after this amount has ionized $= 0.050 - 5.7 \times 10^{-6}$.

Then, letting Y = the amount of NH_4Cl to be added and assuming that the NH_4Cl is completely ionized, there will be an additional Y moles NH_4^+ in solution, so

$$\text{concentration of } OH^- = 5.7 \times 10^{-6}$$

$$\text{concentration of } NH_4^+ = 5.7 \times 10^{-6} + Y$$

$$\text{concentration of } NH_4OH = 0.050 - 5.7 \times 10^{-6}$$

Then, $K_i = 1.8 \times 10^{-5} = \dfrac{[NH_4^+] \ [OH^-]}{[NH_4OH]}$

$$= \frac{(5.7 \times 10^{-6} + Y) \ (5.7 \times 10^{-6})}{(0.050 - 5.7 \times 10^{-6})}$$

and since 5.7×10^{-6} is very small in comparison with 0.050 we can neglect it in the denominator. We can also see by observation that Y will be fairly large so we can neglect 5.7×10^{-6} which is added to it, or

$$1.8 \times 10^{-5} = \frac{(Y) \ (5.7 \times 10^{-6})}{0.050} = \frac{5.7 \times 10^{-6} \ Y}{5.0 \times 10^{-2}}$$

and $Y = \dfrac{1.8 \times 10^{-5} \times 5.0 \times 10^{-2}}{5.7 \times 10^{-6}} = 0.16$ moles per liter

so that 0.16 moles per liter NH_4Cl must be added to prevent the $Mg(OH)_2$ precipitate from forming.

Problem Assignment

Calculate the solubility product constants for the following compounds whose solubilities are:

22-1. $PbCrO_4$ 1.3×10^{-7} moles per liter . . . . (1.7×10^{-14})

22-2. $AgOH$ 1.5×10^{-2} grams per liter . . . . (1.4×10^{-8})

22-3. PbI_2 1.3×10^{-3} moles per liter . . . . (8.8×10^{-9})

22-4. $Fe(OH_3)$ 4.8×10^{-8} grams per liter . . . . (1.1×10^{-36})

22-5. MgF_2 7.4×10^{-2} grams per liter

22-6. $CuBr$ 7.3×10^{-5} moles per liter

22-7. $Sn(OH)_2$ 5.0×10^{-8} grams per liter

22-8. $Al(OH)_3$ 2.0×10^{-7} grams per liter

22-9. $Ca_3(PO_4)_2$ 7.2×10^{-7} moles per liter

Calculate the solubility in moles per liter and in grams per liter for the following compounds:

K_{SP}

22-10. $CdCO_3$ 2.5×10^{-14} .... $(1.6 \times 10^{-7}$ mol/L, 2.8×10^{-5} g/L)

22-11. $Cu(OH)_2$ 5.6×10^{-20} .... $(2.4 \times 10^{-7}$ mol/L, 2.4×10^{-5} g/L)

22-12. BaF_2 1.7×10^{-6} .... $(7.5 \times 10^{-3}$ mol/L, 1.3 g/L)

22-13. $BaSO_4$ 1.10×10^{-10}

22-14. Ag_3PO_4 1.8×10^{-18}

22-15. $Sn(OH)_4$ 1.0×10^{-56}

22-16. $KClO_4$ 1.07×10^{-2}

22-17. Ag_2CrO_4 1.2×10^{-12}

22-18. To one liter of 0.500 M $Pb(NO_3)_2$ is added one liter of 0.500 M NaCl. Will $PbCl_2$ precipitate? $K_{SP} = 1.7 \times 10^{-5}$.

22-19. To 500 mL of 0.0020 M $Mg(NO_3)_2$ is added solid NaOH until the pH reaches 9.65. Will $Mg(OH_2)$ precipitate? $K_{SP} = 1.6 \times 10^{-12}$.

22-20. 1.00 g $BaCl_2$ are added to 400 mL of 0.0050 M H_2SO_4 solution. Will $BaSO_4$ precipitate? $K_{SP} = 1.1 \times 10^{-10}$.

22-21. What is the concentration of Fe^{3+} ions in a saturated solution of $Fe(OH)_3$ when the pH is adjusted to 10.40? $K_{SP} = 1.1 \times 10^{-36}$.

22-22. At what pH will $Zn(OH)_2$ begin to precipitate from a 0.0500 M solution of $ZnCl_2$? $K_{SP} = 3.0 \times 10^{-16}$.

22-23. How many grams of AgCl will dissolve in 10.0 liters of water? $K_{SP} = 1.80 \times 10^{-10}$. How much will dissolve in 10.0 liters of 0.100 M KCl?

22-24. How many grams of $Ni(OH)_2)$ will dissolve in 2.00 liters of a solution containing 4.00 grams of NaOH? $K_{SP} = 6.0 \times 10^{-16}$.

22-25. A solution contains the following ions in 0.100 M concentration: Cd^{2+}, Cu^{2+}, Co^{2+}, Fe^{2+}, Pb^{2+}, Mn^{2+}, Ni^{2+}, Hg^{2+}, Zn^{2+} and Bi^{3+}. If H_2S is bubbled through the solution so that the S^{2-} concentration is 1.00×10^{-22}, which ions will precipitate as sulfides? K_{SP} are:
CdS, 3.6×10^{-29}; CoS, 5×10^{-22}; CuS, 6×10^{-33}; FeS, 1.0×10^{-19}; PbS, 3×10^{-28}; MnS, 2×16^{-53}; NiS, 3.0×10^{-20}; HgS, 2.0×10^{-53}; ZnS, 2×10^{-25}; Bi_2S_3, 1.6×10^{-72}.

22-26. How many individual Ag^+ ions are present per mL in a saturated solution of AgBr? $K_{SP} = 5.0 \times 10^{-13}$.

22-27. 1.0×10^7 Pb^{2+} ions are present per mL of a saturated solution of PbS. What is the K_{SP} for PbS?

22-28. To 1.00 liter of a saturated solution of $BaCO_3$ is added solid K_2CO_3 so that the final CO_3^{2-} concentration is 0.0100 M. Calculate the concentration of the Ba^{2+} ions left and the moles of $BaCO_3$ that precipitated. $K_{SP} = 5.0 \times 10^9$.

22-29. A solution contains 40 mg $CdCl_2$ per mL. At what pH will $Cd(OH)_2$ begin to precipitate? $K_{SP} = 1.2 \times 10^{-14}$.

22-30. If the pH in Problem 22-29 is changed to 12.60, what will be the concentration of Cd^{2+} ions left?

Chapter 23

ELECTROCHEMISTRY

I. FARADAY'S LAW

If we set up a cell consisting of two Cu electrodes immersed in a solution of $CuSO_4$ and connect these electrodes, one to each terminal of a battery, we will be duplicating the experiment carried out by Michael Faraday in 1832.

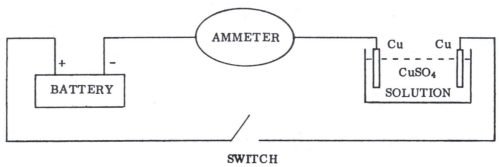

The electrode connected to the negative terminal of the battery is also negative and the one connected to the positive terminal is positive. At the negative terminal, electrons are being "pumped in" from the battery. Cu metal itself cannot gain electrons to relieve the stress but Cu^{2+} can and does, so that

$$Cu^{2+} + 2e^- \rightarrow Cu$$

This is a reduction reaction. This electrode, the one at which reduction occurs, is called the *cathode*.

At the positive electrode, the battery is pulling electrons away and the only thing that can supply them is the Cu, so

$$Cu \rightarrow Cu^{2+} + 2e^-$$

This is oxidation. This electrode, the one at which oxidation occurs, is called the *anode*. Thus we see that Cu is plating out at the cathode and is being dissolved from the anode.

Suppose we weigh the cathode and find it to be 50.000 grams. Then we will place it in the circuit and close the switch, observing the time and noting the current as indicated by the ammeter (10 amps in this case). At the end of 1 hour we open the switch, dry the cathode, and weigh it again. Now it weighs 61.850 grams.

The amount of Cu deposited is 61.850 – 50.000 or 11.850 grams. Now we replace the cathode, close the switch and allow the cell to operate for another hour. This time the cathode weighs 73.700 grams, so that the total amount of Cu plated out is 73.700 – 50.000 or 23.700 grams. NOTE: The amount of Cu deposited in 2 hours is twice that deposited in 1 hour so that the amount of Cu deposited is directly proportional to the time the current flows (since all other factors are constant). Faraday stated this as follows: THE WEIGHT OF A SUBSTANCE PRODUCED AT AN ELECTRODE IN AN ELECTROLYTIC CELL IS DIRECTLY PROPORTIONAL TO THE QUANTITY OF ELECTRICITY PASSING THROUGH THE CELL. The unit of quantity for electricity is the coulomb. A coulomb is the amount of electricity which flows through a wire in one second when the current is one ampere. That is, coulombs = (number of amps) × (number of seconds).

Consider the following reactions:

$$Ag^+ \; + e^- \;\; \rightarrow Ag$$
$$Cu^{2+} \; + 2\,e^- \rightarrow Cu$$
$$Al^{3+} \; + 3\,e^- \rightarrow Al$$

These equations tell us that:

$\qquad$ 1 mole of electrons produces one mole of silver

$\qquad$ 1 mole of electrons produces 1/2 mole copper

$\qquad$ 1 mole of electrons produces 1/3 mole aluminum

The quantity of electricity associated with one mole of electrons is call a FARADAY. One Faraday is equivalent to 96,500 coulombs.

Example No. 23-1. 3.00 Faradays of electricity are passed through molten NaCl. How many grams of sodium will be produced? How many grams of chlorine? How many liters of chlorine at STP?

First, we write the half reactions involved.

$$Na^+ + e^- \rightarrow Na$$

$$2\ Cl^- - 2e^- \rightarrow Cl_2$$

We know that one Faraday is equivalent to one mole of electrons, so we will first change Faradays to moles of electrons; then moles of electrons to moles of sodium; and finally moles of sodium to grams of sodium.

Thus:

$$3\ \text{Faradays} \times \frac{1\ \text{mole electrons}}{1\ \text{Faraday}} = 3\ \text{moles electrons}$$

$$3\ \text{moles electrons} \times \frac{1\ \text{mole Na}}{1\ \text{mole electrons}} = 3\ \text{moles Na}$$

$$3\ \text{moles Na} \times \frac{23.0\ \text{g Na}}{1\ \text{mole Na}} = 69.0\ \text{g Na}$$

Combining all this into one step,

$$3\ \text{Faradays} \times \frac{1\text{mole } e^-}{1\ \text{Faraday}} \times 1\ \frac{\text{mole Na}}{1\ \text{mole } e^-} \times \frac{23.0\ \text{g Na}}{1\ \text{mole Na}} = 69\ \text{g Na}$$

$$\text{Faradays} \quad \rightarrow \quad \text{moles } e^- \rightarrow \quad \text{moles Na} \rightarrow \quad \text{g Na}$$

Likewise, for the chlorine,

$$3\ \text{Faradays} \times \frac{1\ \text{mole } e^-}{1\ \text{Faraday}} \times \frac{1\ \text{mole Cl}_2}{2\ \text{moles } e^-} \times \frac{71.0\ \text{g Cl}_2}{1\ \text{mole Cl}_2} = 106\ \text{g Cl}_2$$

To find the volume of Cl_2 at STP, we recall that 1 mole of gas (at STP) occupies 22.4 liters, so

$$3\ \text{Faradays} \times \frac{1\ \text{mole } e^-}{1\ \text{Faraday}} \times \frac{1\ \text{mole Cl}_2}{2\ \text{moles } e^-} \times \frac{22.4\ \text{L Cl}_2}{1\ \text{mole Cl}_2} = 33.6\ \text{L Cl}_2\ \text{at STP}$$

Example No. 23-2. How much silver will be plated from a solution of silver nitrate by a current of 5.00 amperes flowing for 10.0 hours?

First, we must calculate how many coulombs of electricity are being used.

$$\text{Coulombs} = \text{amperes} \times \text{seconds}$$

$$= 5.00\ \text{amperes} \times 10.0\ \text{hours} \times 3600\ \frac{\text{seconds}}{\text{hour}} = 180{,}000\ \text{coulombs}.$$

Then, changing coulombs to Faradays (1 Faraday = 96,500 coulombs) to moles e^- to moles Ag (Ag^+ + e^- → Ag) to grams Ag, we have

$$180,000 \; \text{coulombs} \times \frac{1 \; \text{Faraday}}{96,500 \; \text{coulombs}} \times \frac{1 \; \text{mol } e^-}{1 \; \text{Faraday}} \times \frac{1 \; \text{mol Ag}}{1 \; \text{mol } e^-} \times \frac{108 \; \text{g Ag}}{1 \; \text{mol Ag}} = 201 \; \text{g Ag}$$

Example No. 23-3. How long will it take to produce 50.0 grams Al from the molten oxide using a current of 25.0 amperes?

$$Al^{3+} + 3e^- \rightarrow Al$$

First, we will change g Al to moles Al. Then, moles Al to moles e^- to Faradays and finally to coulombs, or

$$50.0 \; \text{g Al} \times \frac{1 \; \text{mol Al}}{27.0 \; \text{g Al}} \times \frac{3 \; \text{mol } e^-}{1 \; \text{mol Al}} \times \frac{1 \; \text{Faraday}}{1 \; \text{mol } e^-} \times \frac{96,500 \; \text{coulombs}}{1 \; \text{Faraday}} = 536,000 \; \text{coulombs}$$

Next, since coulombs = amps × seconds,

$$\text{seconds} = \frac{\text{coulombs}}{\text{amps}} = \frac{536,000 \; \text{coulombs}}{25.0 \; \text{amps}} = 21,400 \; \text{seconds}$$

$$21,400 \; \text{seconds} \times \frac{1 \; \text{hour}}{3600 \; \text{seconds}} = 5.94 \; \text{hours}$$

COST PROBLEMS

Many times we are interested in the cost of a plating operation. Electricity is usually sold by the kilowatt hour, or KWH, which is a product of the number of kilowatts and the number of hours.

1 kilowatt = 1000 watts

1 watt   = 1 volt × 1 amp.

Example No. 23-4. How much will it cost, at 8.0¢ per KWH, to plate Cu from a solution of $CuSO_4$ using a current of 5.00 amperes at 2.0 volts if the cell is in operation for 24.0 hours? How much Cu will be produced? (Cu^{2+} + 2 e^- → Cu).

First, we must calculate the number of watts used.

Watts = volts × amps = 2.0 volts × 5.00 amps = 10 watts = 0.010 KW

The number of KWH is 0.010 KW × 24.0 hours = 0.24 KWH. At 8.0¢ per KWH, the cost is 0.24 KWH × 8.0¢ per KWH = 1.92¢.

To calculate the amount of Cu produced.

$$24.0 \; \text{hours} = 24.0 \; \text{hours} \times \frac{3600 \; \text{seconds}}{\text{hour}} = 86,400 \; \text{seconds}$$

Coulombs = amperes × seconds = 5.00 amperes × 86,400 seconds = 432,000 coulombs.

$$432,000 \; \text{coulombs} \times \frac{1 \; \text{Faraday}}{96,500 \; \text{coulombs}} \times \frac{1 \; \text{mol } e^-}{1 \; \text{Faraday}} \times \frac{1 \; \text{mol Cu}}{2 \; \text{mol } e^-} \times \frac{63.6 \; \text{g Cu}}{1 \; \text{mol Cu}} = 142 \; \text{g Cu}$$

Problem Assignment

23-1. How many grams of Cr will be plated out of a $CrCl_3$ solution by a current of 10.0 amperes flowing for 6.00 hours? . . . . (38.8 g)

23-2. How much Al will be produced per 8.00 hour day by the electrolysis of Al_2O_3 in molten cryolite using a current of 25.0 amperes? . . . . (67.2 g)

23-3. 83.8 g of a metal are plated per hour by a current of 40.0 amps. If the reaction is M^{2+} + 2 e → M, what is the molecular mass of the metal? . . . . (112)

23-4. How long will it take to plate 1.00 pound of Mg from molten $MgCl_2$ using a current of 50.0 amps? . . . . (20.0 hours)

23-5. How much will it cost, at 6.00¢/KWH to plate 10.0 kg Cr from a $CrCl_3$ solution using a 6.00 volt, 100 amp power supply? . . . . ($5.57)

23-6. 10.0 amps flow through molten NaCl for one day.
    a. How many grams of Na will be produced?
    b. How many moles Na?
    c. How many grams Cl_2?
    d. How many moles Cl_2?
    e. How many liters Cl_2 at 600°C and one atmosphere pressure?

23-7. In order to plate out 200 g Cu per hour from a $CuSO_4$ solution, how many Faradays are necessary? How many coulombs? How many amperes?

23-8. A rectifier supplies 10.0 amps at 3.00 volts. It is connected in series to a solution of (a) $AgNO_3$, (b) $CuSO_4$, (c) dilute HCl and allowed to run for one day. How much Ag will be produced? How much Cu? What volume H_2 and O_2 at STP? What will it cost to operate these cells at 3.00¢/KWH?

23-9. Two cells are connected so that the same amount of current flows through each. One cell contains $CuSO_4$ and the other $AgNO_3$. If 250 g of Cu are plated out, how much Ag will be plated?

23-10. How much would it cost to plate 4.00 lb. Al at 8.00¢/KWH using a 4.00 volt power supply?

23-11. How long will it take to plate out all the Ni from 1500 mL of 0.300 M $NiSO_4$ solution, using 10.0 amps?

23-12. A 5.00 cm × 5.00 cm thin piece of Copper is placed in a $AgNO_3$ solution and Ag plated on it for 4.00 hours using 20.0 amps. How thick will the plate be, assuming equal plating on both sides? Density of Ag = 10.5 g/cm^3.

23-13. What current is necessary to produce 200 liters H_2 (at STP) per hour from a dilute H_2SO_4 solution? What volume of O_2 will be produced in the same period of time?

II. ENERGIES FROM STANDARD REDUCTION POTENTIALS

An oxidation-reduction reaction may be considered as being the sum of two half-reactions—one involving oxidation and the other reduction. Each half-reaction may be assigned an electromotive force (voltage) or $\varepsilon°$ value.

The $\varepsilon°$ values for all reduction half-cell reactions are determined by comparison with a standard Hydrogen electrode consisting of 1 M H^+ and 1 atmosphere pressure of H_2 gas at 25°C. Arbitrarily, the standard Hydrogen electrode is assigned an $\varepsilon°$ value of 0.00 volts.

Table 23-1 lists the $\varepsilon°$ values of several half-cell reactions.

Table 23-1. Standard Reduction Potentials at 25°C

	$\varepsilon°$		$\varepsilon°$
$F_2 + 2e^- \rightarrow 2F^-$	2.87	$Pb^{2+} + 2e^- \rightarrow Pb$	-0.13
$MnO_4^- + 8H^+ + 5e^- \rightarrow Mn^{2+} + 4H_2O$	1.51	$Sn^{2+} + 2e^- \rightarrow Sn$	-0.14
$Cl_2 + 2e^- \rightarrow 2Cl^-$	1.36	$Ni^{2+} + 2e^- \rightarrow Ni$	-0.28
$Cr_2O_7^{2-} + 14H^+ + 6e^- \rightarrow 2Cr^{3+} + 7H_2O$	1.23	$Co^{2+} + 2e^- \rightarrow Co$	-0.28
$O_2 + 4H^+ + 4e^- \rightarrow 2H_2O$	1.20	$Cd^{2+} + 2e^- \rightarrow Cd$	-0.40
$2IO_3^- + 12H^+ + 10e^- \rightarrow I_2 + 6H_2O$	1.19	$Cr^{3+} + e^- \rightarrow Cr^{2+}$	-0.41
$Br_{2(l)} + 2e^- \rightarrow 2Br^-$	1.06	$Fe^{2+} + 2e^- \rightarrow Fe$	-0.44
$Hg^{2+} + 2e^- \rightarrow Hg$	0.85	$2CO_2 + 2H^+ + 2e^- \rightarrow H_2C_2O_4$	-0.49
$Ag^+ + e^- \rightarrow Ag$	0.80	$S + 2e^- \rightarrow S^{2-}$	-0.51
$Fe^{3+} + e^- \rightarrow Fe^{2+}$	0.77	$Cr^{3+} + 3e^- \rightarrow Cr$	-0.74
$O_2 + 2H^+ + 2e^- \rightarrow H_2O_2$	0.68	$Zn^{2+} + 2e^- \rightarrow Zn$	-0.76
$H_3AsO_4 + 2H^+ + 2e^- \rightarrow HAsO_2 + 2H_2O$	0.56	$2SO_3^{2-} + 2H_2O + 2e^- \rightarrow S_2O_4^{2-} + 4OH^-$	-1.12
$I_2 + 2e^- \rightarrow 2I^-$	0.54	$Mn^{2+} + 2e^- \rightarrow Mn$	-1.18
$O_2 + 2H_2O + 4e^- \rightarrow 4OH^-$	0.40	$Al^{3+} + 3e^- \rightarrow Al$	-1.66
$Cu^{2+} + 2e^- \rightarrow Cu$	0.34	$Mg^{2+} + 2e^- \rightarrow Mg$	-2.37
$SO_4^{2-} + 4H^+ + 2e^- \rightarrow H_2SO_3 + H_2O$	0.20	$Na^+ + e^- \rightarrow Na$	-2.71
$S_4O_6^{2-} + 2e^- \rightarrow 2S_2O_3^{2-}$	0.17	$Ca^{2+} + 2e^- \rightarrow Ca$	-2.87
$Cu^{2+} + e^- \rightarrow Cu^+$	0.15	$Ba^{2+} + 2e^- \rightarrow Ba$	-2.90
$Sn^{4+} + 2e^- \rightarrow Sn^{2+}$	0.15	$K^+ + e^- \rightarrow K$	-2.92
$2H^+ + 2e^- \rightarrow H_2$	0.00	$Li^+ + e^- \rightarrow Li$	-3.05

Example No. 23-5.   Calculate the $\varepsilon°$ value for the reaction.

Cd + Cu^{2+} → Cd^{2+} + Cu, assuming 1 M concentrations.

The above reaction may be written as the sum of two half-reactions:

Cd → Cd^{2+} + $2e^-$ (1)

Cu^{2+} + $2e^-$ → Cu   (2)

The voltage for reaction (1), which is the reverse of the reaction listed in the table, is +0.40 volts. Note that when a reaction is the reverse of the one in the table, the sign of the $\varepsilon°$ value must also be reversed. The voltage for reaction (2), as given in the table, is +0.34 volts.

Thus, the above two half-reactions may be written as

Cd → Cd^{2+} + $2e^-$ +0.40 volts

Cu^{2+} + $2e^-$ → Cu   +0.34 volts

The overall reaction, which is the sum of the above two half-reactions, has an $\varepsilon°$ value equal to the sum of the $\varepsilon°$ values of the half-reactions

Cd → Cd^{2+} + $2e^-$ +0.40 volts

Cu^{2+} + $2e^-$ → Cu   +0.34 volts

—————————————————————————

Cd + Cu^{2+} → Cd^{2+} + Cu   +0.74 volts

Example No. 23-6.   What is the emf of a cell consisting of an iron and lead electrode, each immersed in a solution of its salt, and separated by a membrane?

Fe + Pb^{2+} → Fe^{2+} + Pb (assume 1 M concentrations)

If we write the two half-reactions and their corresponding $\varepsilon°$ values, we have

Fe → Fe^{2+} + $2e^-$ +0.44 volts

Pb^{2+} + $2e^-$ → Pb   −0.13 volts

—————————————————————————

Fe + Pb^{2+} → Fe^{2+} + Pb   +0.31 volts

Example No. 23-7.   Will the following reaction take place?

MnO_4^- + Fe^{2+} + H^+ → Mn^{2+} + Fe^{3+} + H_2O (1 M solutions)

First, we break this equation into two half-reactions

MnO_4^- + H^+ → Mn^{2+} + H_2O and

Fe^{2+} → Fe^{3+}

Then, from the table of $\varepsilon°$ values we see that

MnO_4^- + H^+ → Mn^{2+} + H_2O has an emf of 1.51 volts.

Fe^{2+} → Fe^{3+} has an emf of −0.77 volts.

Then, since we add the half-reactions to get the overall reaction, we also add $\varepsilon°$ values or

$MnO_4^- + H^+ \rightarrow Mn^{2+} + H_2O$	1.51 volts
$Fe^{2+} \rightarrow Fe^{3+}$	-0.77 volts
$MnO_4^- + Fe^{2+} + H^+ \rightarrow Mn^{2+} + Fe^{3+} + H_2O$	$+0.74$ volts

Since the voltage is positive this reaction will occur. (Note that this reaction is not balanced.)

Example No. 23-8.   Predict whether the following reaction will occur.

$$S_4O_6^{2-} + I^- \rightarrow S_2O_3^{2-} + I_2 \quad (1\ M\ \text{solutions})$$

First, we write the two half-reactions, or

$S_4O_6^{2-} \rightarrow S_2O_3^{2-}$, and

$I^- \rightarrow I_2$

From the table of $\varepsilon°$ values we see that for

$S_4O_6^{2-} \rightarrow S_2O_3^{2-}$ the emf is $+0.17$.

$I^- \rightarrow I_2$, the emf is -0.54.

Then, when we add the half-reactions, we get

$S_4O_6^{2-} \rightarrow S_2O_3^{2-}$	$+0.17$ volts
$I^- \rightarrow I_2$	-0.54 volts
$S_4O_6^{2-} + I^- \rightarrow S_2O_3^{2-} + I_2$	-0.37 volts

where a negative voltage indicates that the reaction will not work. (Actually, a negative value indicates that the reaction will take place in the reverse direction.) However, a positive voltage is not the only criterion as to whether a reaction will proceed or not. There must also be oxidation and reduction.

Predict (from $\varepsilon°$ values) whether the following reactions (unbalanced) will occur. If a reaction does occur, give the emf; if a reaction does not occur, write NO REACTION.

23-14. $Cd^{2+} + Zn \rightarrow Zn^{2+} + Cd$

23-15. $Mn^{2+} + Mg \rightarrow Mg^{2+} + Mn$

23-16. $Cr + Cd \rightarrow Cr^{3+} + Cd^{2+}$

23-17. $Pb + Fe^{2+} \rightarrow Pb^{2+} + Fe^{3+}$

23-18. $Fe^{2+} + Sn^{2+} \rightarrow Fe^{3+} + Sn$

23-19. $Cr^{3+} + CO_2 + H_2O \rightarrow Cr_2O_7^{2-} + H_2C_2O_4 + H^+$

23-20. $I_2 + S_2O_3^{2-} \rightarrow I^- + S_4O_6^{2-}$

23-21. $MnO_4^- + HAsO_2 + H^+ + H_2O \rightarrow H_3AsO_4 + Mn^{2+}$

23-22. $Fe^{3+} + S^{2-} \rightarrow Fe^{2+} + S$

23-23. $Cl_2 + I^- \rightarrow Cl^- + I_2$

23-24. $Cd + Hg^{2+} \rightarrow Cd^{2+} + Hg$

23-25. $MnO_4^- + Fe^{2+} + H^+ \rightarrow Mn^{2+} + Fe^{3+} + H_2O$

III.   THE NERNST EQUATION

As was shown in Example No. 23-5, the electromotive force for the reaction

$$Cd + Cu^{2+} \rightarrow Cd^{2+} + Cu$$

is 0.74 volts. However, this voltage will be produced only if the concentrations of Cu^{2+} and Cd^{2+} are $1M$. If the concentrations are not 1 M, then the emf of the cell will differ from that predicted by the use of $\varepsilon°$ values.

The emf of a cell whose net reaction is $aX + bY \rightarrow cZ + dS$ and whose concentrations are not 1 M may be calculated by means of the Nernst equation

$$\varepsilon = \varepsilon° - \frac{0.0592}{n}\log \frac{[Z]^c [S]^d}{[X]^a [Y]^b}$$

where n is the number of electrons transferred in the reaction (temperature remaining constant at 25°C).

The above equation may be written as

$$\varepsilon = \varepsilon° - \frac{0.0592}{n}\log \frac{(concentration\ of\ products)}{(concentration\ of\ reactants)}$$

with each being raised to the appropriate power.

Example No. 23-9.   Calculate the emf of a cell consisting of a cadmium rod in 0.10 M $CdSO_4$ solution and a copper bar in 0.010 M $CuSO_4$ solution.

$$Cd + Cu^{2+} \rightarrow Cd^{2+} + Cu$$

We have already calculated $\varepsilon°$ to be +0.74 volts (at 25°C) if the concentrations are 1 M.

Then $\varepsilon = \varepsilon° - \dfrac{0.0592}{n}\log \dfrac{[Cd^{+2}]}{[Cu^{+2}]}$ where n (number of electrons transferred) = 2.

Note that metallic cadmium and copper are not included in the concentration term since their concentrations are constant.

$$\varepsilon = 0.74 - \frac{0.0592}{n}\log \frac{[0.10]}{[0.010]}$$

$$= 0.74 - \frac{0.0592}{2}\log 10 = 0.71 \text{ volts.}$$

Example No. 23-10.   What is the emf of a cell whose net reaction is

$$2\ Al + 3\ Cl_2 \rightarrow 2\ Al^{3+} + 6\ Cl^- \text{ if } Al^{3+} \text{ concentration is}$$

0.10 M and Cl^- concentration is 0.020 M?

First, the net reaction may be written as 2 half-reactions, each with a certain standard emf as found in the table.

$$Al \rightarrow Al^{3+} + 3e \qquad + 1.66 \text{ volts}$$

$$Cl_2 + 2e \rightarrow 2\ Cl^- \qquad + 1.36 \text{ volts}$$

Note that the first half-reaction is the reverse of the one in the table so that the sign of the emf is reversed.

Before the above two half-reactions may be added, the number of electrons lost must equal the number gained. Thus, upon multiplying the first half-reaction by 2 and the second by 3, the number of electrons lost and gained become equal.

$2\ Al \rightarrow 2\ Al^{3+} + 6e^-$	$+\ 1.66$ volts
$3\ Cl_2 + 6e^- \rightarrow 6\ Cl^-$	$+\ 1.36$ volts
$2\ Al + 3\ Cl_2 \rightarrow 2\ Al^{3+} + 6\ Cl^-$	$+\ 3.02$ volts

That is, the emf produced by such a cell, if ion concentrations are 1 M, is 3.07 volts.

Hence, since ion concentrations are not 1 M,

$$\varepsilon = \varepsilon° - \frac{0.0592}{n}\log\frac{[\text{concentrations of products}]}{[\text{concentrations of reactants}]}$$

$$\varepsilon = 3.02 - \frac{0.0592}{n}\log [Al^{3+}]^2\ [Cl^-]^6$$

It should be noted that the number of electrons being transferred in the net equation $2\ Al + 3\ Cl_2 \rightarrow 2\ Al^{3+} + 6\ Cl^-$ is six. Also the concentration of Al + Cl_2 are not included in the above equation since they are constant (not ionic).

Thus, $\varepsilon = 3.02 - \dfrac{0.0592}{6}\log (0.10)^2\ (0.020)^6 = 3.14$ volts

Example No. 23.11. A concentration cell consists of a silver electrode in 0.10 M silver nitrate solution and another silver electrode in 1.00 M silver nitrate solution. What emf will such a cell produce?

A concentration cell consists of two solutions of different concentrations separated by a salt bridge with an electrode in each solution. The electrode in the dilute solution will tend to lose electrons (become oxidized) while the ions in the more concentrated solution will tend to gain electrons (become reduced). The net half-reactions, and emfs, are,

Dilute (0.10 M) solution	$Ag^+ \rightarrow Ag + e^-$	-0.80 volts
Concentrated (1.0 M solution	$Ag^+ + e^- \rightarrow Ag$	$+0.80$ volts
Net reaction $Ag + Ag^+ \rightarrow Ag^+ + Ag$		0.00 volts

Thus, the emf produced by the net reaction in 1 M solutions should be zero. However, since the concentrations are not 1 M,

$$\varepsilon = \varepsilon° - \frac{0.0592}{n}\log\frac{(Ag^+)\ (\text{dilute solution})}{(Ag^+)\ (\text{concentrated solution})}$$

$$= 0 - \frac{0.0592}{1}\log\frac{0.10}{1} = 0.0592\text{volts}$$

Example No. 23-12. A cell consists of a copper electrode immersed in 1 M $CuSO_4$ solution and a hydrogen electrode (Platinum electrode exposed to H_2 gas at 1 atm pressure) immersed in an acidic solution. Calculate the pH of that solution if the cell produces 0.46 volts.

The net reaction is

$$Cu^{2+} + H_2 \rightarrow Cu + 2\,H^+$$

The emf of this reaction, at 1 M concentrations, is

$Cu^{2+} + 2e \rightarrow Cu$	+0.34
$H_2 \rightarrow 2\,H^+ + 2e$	0.00

$$Cu^{2+} + H_2 \rightarrow Cu + 2\,H^+ \qquad +0.34$$

$$\varepsilon = \varepsilon° - \frac{0.0592}{n}\log \frac{(\text{concentration of products})}{(\text{concentration of reactants})}$$

$$0.46 = 0.34 - \frac{0.0592}{n}\log \frac{(Cu)\ (H^+)^2}{(Cu^{2+})\ (H_2)}$$

and since concentration of Cu (a solid) and H_2 (a gas at 1 atm pressure) are constant

$$0.46 = 0.34 - \frac{0.0592}{n}\log \frac{(H^+)^2}{(Cu^{2+})}$$

$$0.46 = 0.34 - \frac{0.0592}{2}\log \frac{(H^+)^2}{(1)}$$

$$0.46 = 0.34 - \frac{0.0592}{2}\log (H^+)^2$$

Next, since $\log (H^+)^2 = 2 \log (H^+)$,

$$0.46 = 0.34 - \frac{0.0592}{2}(2)\log (H^+)$$

$$0.46 = 0.34 - 0.0592 \log (H^+)$$

$$0.0592 \log (H^+) = -0.12$$

$$\log (H^+) = -2.0$$

However, since pH is defined as $-\log (H^+)$

$$\text{pH of solution} = 2.0$$

Example No. 23-13. Calculate the emf of a zinc-chromium cell if

(a) concentration of both $Zn^{2+} + Cr^{3+}$ are 1 M

(b) concentrations of Zn^{2+} and Cr^{3+} are 0.10 M and 1.0 M respectively

(c) concentrations of Zn^{2+} and Cr^{3+} are 1.0 M and 0.010 M respectively

$$3\,Zn + 2\,Cr^{3+} \rightarrow 3\,Zn^{2+} + 2\,Cr$$

a) If both ion concentrations are 1 M,

3 Zn	$\rightarrow$ 3 Zn^{2+} + 6e$^-$	+0.76 volts
2 Cr^{3+} + 6e$^-$	$\rightarrow$ 2 Cr	−0.74 volts

$$3 \text{ Zn} + 2 \text{ Cr}^{3+} \rightarrow 3 \text{ Zn}^{2+} + 2 \text{ Cr} \qquad +0.02 \text{ volts}$$

b) If Zn^{2+} and Cr^{3+} concentrations are 0.10 M and 1.0 M respectively

$$\varepsilon = 0.02 - \frac{0.0592}{6} \log \frac{(0.10)^3}{(1)^2} = 0.02 + 0.03 = 0.05 \text{ volts}$$

c) If concentrations of Zn^{2+} + Cu^{3+} are 1.0 M and 0.010 M

$$\varepsilon = 0.02 - \frac{0.0592}{6} \log \frac{(1.0)^3}{(0.010)^2} = 0.02 - 0.04 = -0.02 \text{ volts}$$

which indicates that under these conditions the reaction proceeds in the reverse direction (from right to left).

Equilibrium Constants and Emf of Cells

As a cell produces an emf, the concentration of the products increases, the concentration of the reactants decreases, and the voltage steadily drops.

Eventually the voltage of the cell will become zero and there will be no further chemical reaction. At this time the concentrations of products and reactants remain constant indefinitely. That is, when the emf equals zero, the equilibrium condition has been reached.

For the reaction $aX = bY \rightleftharpoons cZ + dS$

$$\varepsilon = \varepsilon° - \frac{0.0592}{n} \log \frac{(Z)^c \ (S)^d}{(X)^a \ (Y)^b}$$

At equilibrium, $\varepsilon = 0$, so

$$0 = \varepsilon° - \frac{0.0592}{n} \log \frac{(Z)^c \ (S)^d}{(X)^a \ (Y)^b}$$

However, $\dfrac{(Z)^c \ (S)^d}{(X)^a \ (Y)^b} = K_{eq}$ so

$$\varepsilon° - \frac{0.0592}{n} \log K_{eq} \quad \text{and}$$

$$\log K_{eq} = \frac{n \ \varepsilon°}{0.0592}$$

Example No. 23-14. Calculate the value of the equilibrium constant for the reaction:

$$Cr_2O_7^{2-} + Br^- + H^+ \rightarrow Cr^{3+} + Br_2 + H_2O$$

Writing the above net reaction as two half-reactions we have:

$$Cr_2O_7^{2-} + 14\ H^+ + 6\ e^- \rightarrow 2\ Cr^{3+} + 7\ H_2O \quad +1.33 \text{ volts}$$

$$2\ Br^- \qquad\qquad\qquad \rightarrow Br_2 + 2e^- \qquad\qquad -1.06 \text{ volts}$$

Multiplying the second half-reaction by three to equal the loss and gain:

$$Cr_2O_7^{2-} + 14\ H^+ + 6\ e^- \rightarrow 2\ Cr^{3+} + 7\ H_2O \qquad\qquad +1.33 \text{ volts}$$

$$2\ Br^- \qquad\qquad\qquad \rightarrow Br_2 + 2e^- \qquad\qquad\qquad -1.06 \text{ volts}$$

$$Cr_2O_7^{2-} + 6\ Br^- + 14\ H^+ \rightarrow 2\ Cr^{3+} + 3\ Br_2 + 7\ H_2O \qquad +0.27 \text{ volts}$$

Then

$$\log K_{eq} = \frac{n\ \mathcal{E}^\circ}{0.0592} = \frac{6(0.27)}{0.0592}$$

$$\log K_{eq} = 27.36$$

$$K_{eq} = 10^{27.36} = 2.3 \times 10^{27}$$

Example No. 23-15. A piece of chromium is placed in a 1.0 M Zn^{2+} solution. What is the equilibrium concentration of Cr^{3+}?

$$3\ Zn^{2+} + 2\ Cr \rightarrow 3\ Zn + 2\ Cr^{3+}$$

First, to find $\mathcal{E}^\circ$ for the reaction,

$$3\ Zn^{2+} + 6e^- \rightarrow 3\ Zn \qquad\qquad -0.76 \text{ volts}$$

$$2\ Cr \qquad\qquad \rightarrow 2\ Cr^{3+} + 6e^- \qquad +0.74 \text{ volts}$$

$$3\ Zn^{2+} + 2\ Cr \rightarrow 3\ Zn + 2\ Cr^{3+} \qquad -0.02 \text{ volts}$$

Then, $\log K_{eq} = n\ \mathcal{E}^\circ = \dfrac{6(-0.02)}{0.0592} = -2.03$

$K_{eq} = 9.3 \times 10^{-3}$

Then, since K_{eq} for the above reactions $= \dfrac{(Cr^{3+})^2}{(Zn^{2+})^3}$ (concentrations of Cr and Zn being constant),

$$9.3 \times 10^{-3} = \frac{(Cr^{3+})^2}{(Zn^{2+})^3} = \frac{(Cr^{3+})^2}{(1.0)^3}$$

$$(Cr^{3+}) = \sqrt{9.3 \times 10^{-3}} = 9.6 \times 10^{-2}\ M$$

Calculate the emf of each of the following cells:

23-26. $Zn|Zn^{2+}$ (0.10 M) ‖ Cu^{2+} (1.0 M) | Cu .... (1.13 volts)
which indicates a zinc electrode in contact with a 0.10 M zinc solution and a copper electrode
in contact with a 1.0 M Cu^{2+} solution and with the oxidation half-reaction indicated first.

23-27. $Co|Co^{2+}$ (0.001 M) ‖ H^+ (0.1 M) | H_2 (1 atm)

23-28. $Cu|Cu^{2+}$ (0.01 M) ‖ Cu^{2+} (1.0 M) | Cu .... (0.0592 volts)

23-29. $Ni|Ni^{2+}$ (0.10 M) ‖ Sn^{2+} (1 M) |Sn

23-30. H_2 (1 atm) | H^+ (pH = 3) ‖ Ag^+ (0.10 M) | Ag .... (0.92 volts)

Calculate K_{eq} for each of the following reactions:

23-31. $Zn + Cu^{2+} \rightarrow Zn^{2+} + Cu$ .... ($10^{37.2}$)

23-32. $Sn^{2+} + Co \rightarrow Co^{2+} + Sn$

23-33. $Cr_2O_7^{2-} + Cl^- + H^+ \rightarrow Cr^{3+} + Cl_2 + H_2O$ .... ($10^{-3.04}$)

23-34. $Ag + Fe^{3+} \rightarrow Fe^{2+} + Ag^+$

23-35. $Na + F_2 \rightarrow Na^+ + F^-$ .... (10^{189})

23-36. A strip of copper is placed in 0.001 M Co^{2+} solution. What is the equilibrium concentration
of Cu^{2+}?

23-37. A piece of lead is placed in a 0.10 M Sn^{2+} solution. Calculate the equilibrium concentration
of Pb^{2+}? .... (0.21 M)

Chapter 24

NUCLEAR REACTIONS

I. NUCLEAR MASSES

Previously, in the chapter on atomic structure, we assumed the masses of the proton and the neutron to be 1 unit each and that of the electron to be 1/1837 of this unit. Now, before we can begin calculations involving nuclear reactions, we must have more exact information to work with. Thus, in the following table we have more accurate nuclear masses.

Table 24-1. Nuclear Masses of Representative Elements.

	atomic number	mass number	mass*
n	0	1	1.00866
H	1	1	1.00728
	1	2	2.01355
	1	3	3.01550
He	2	3	3.01493
	2	4	4.00150
C	6	11	11.00814
	6	12	11.99671
	6	13	13.00006
	6	14	13.99995
N	7	14	13.99922
O	8	16	15.99052
	8	17	16.99474
	8	18	17.99477
Mg	12	25	24.97925
Al	13	27	26.97439
Cl	17	35	34.95952
Ar	18	40	39.95250
K	19	40	39.95358
Po	84	218	217.9628
Rn	86	222	221.9703
U	92	239	239.0038

---

* Note that the units of nuclear mass may be either amu (1 amu = 1.66×10^{-27} kg) or g/mol. The type of unit selected will depend upon the problem being solved.

Example No. 24-1.   Calculate the mass of an atom of $_6^{12}$C in amu. Also calculate the mass of a mole of $_6^{12}$C. (Mass of an electron = 0.000549.)

To calculate the mass of an atom of $_6^{12}$C we add to the nuclear mass (11.99671), the masses of 6 electrons, so

mass of $_6^{12}$C = 11.99671 + 6(0.000549) = 12.00000 amu.

Likewise, the mass of a mole of $_6^{12}$C = 12.00000 g.

II. BINDING ENERGY

If we consider the molar nuclear mass of the most abundant isotope of helium, $_2^4$He*, and compare it with the sum of the masses of the 2 protons and 2 neutrons that compose it, we arrive at the following results:

2 protons = 2 × 1.00728 = 2.01456 g/mol

2 neutrons = 2 × 1.00866 = 2.01732 g/mol

Sum of the particles   = 4.03188 g/mol

Nuclear mass (from table) = <u>4.00150</u> g/mol

Difference in mass   = <u>0.03038</u> g/mol

If we consider other atoms, we will find similar results, namely that the nuclear mass is less than the sum of the masses of its constituent particles. What happened to the mass that was lost? Mass cannot disappear completely without leaving an equivalent amount of energy (according to Einstein's theory) and so the lost mass is replaced by an equivalent amount of energy. When the nucleus of an atom is formed from a number of particles, energy is liberated. This energy represents the energy that binds the particles together in the nucleus and is called the *binding energy*. The binding energy is also the energy necessary to separate the particles in the nucleus. A large binding energy indicates a stable nucleus while a small one indicates an unstable nucleus.

To calculate the binding energy we use the Einstein equation

$$\Delta E = \Delta m \ c^2$$

where ΔE represents the binding energy, Δm the change in mass, and c the speed of light. ΔE has the unit joules (J) when m is in kg and c in m/sec.

Thus the binding energy for the helium nucleus is:

$\Delta E = \Delta m \ c^2$ where Δm is 0.03038 g/mol or 3.038×10^{-5} kg/mol and c is 2.998×10^8 m/sec.

So, $\Delta E = 3.038 \times 10^{-5}$ kg/mol $\times (2.998 \times 10^8$ m/sec$)^2 = 2.730 \times 10^{12}$ J/mol.

---

*   In the symbol $_2^4$He, the subscript (lower number) indicates the atomic number and the superscript (upper number) the mass number.

Example No. 24-2.   Calculate the energy change, in J/mol, for the following reaction:

$$^{27}_{13}Al + {}^{2}_{1}H \rightarrow {}^{25}_{12}Mg + {}^{4}_{2}He$$

First, we will find the nuclear masses of the products and reactants to see how much has disappeared. From this we can calculate the energy change.

Mass of products:	$^{25}_{12}Mg$	24.97925 g/mol
(from Table 24-1)		
	$^{4}_{2}He$	4.00150
Total		28.98075

Mass of reactants	$^{27}_{13}Al$	26.97439
	$^{2}_{1}H$	2.01355
Total		28.98794 g/mol
Difference:		0.00719 g/mol

Δm = 0.00719 g/mol or 7.19×10^{-6} kg/mol.

Then, $\Delta E = \Delta m\, c^2 = 7.19 \times 10^{-6}$ kg/mol $\times (3.00 \times 10^8$ m/sec$)^2$

$\Delta E = 6.47 \times 10^{11}$ J/mol.

III.   NATURAL RADIOACTIVITY

Naturally radioactive substances may emit alpha particles (α), beta particles (β), gamma rays (γ) or any combination of these. Alpha particles are helium nuclei; beta particles, high speed electrons; and gamma rays are a form of electromagnetic radiation similar to X-rays.

A alpha particle, α, which is a helium nucleus, is indicated by the symbol $^{4}_{2}He$.

A beta particle, β, which is an electron, is indicated by the symbol $^{0}_{-1}e$.

A gamma ray, γ, which is an electromagnetic radiation and not a particle, is indicated merely as γ.

In any nuclear reaction the following rules must be observed.

The sum of atomic numbers (the subscripts) must be the same on both sides of the equation. Also, the sum of the mass numbers (the superscripts) must be the same on both sides of the equation.

$$^{14}_{7}N + {}^{4}_{2}He \rightarrow {}^{17}_{8}O + {}^{1}_{1}H \qquad \begin{matrix}(14 + 4 = 17 + 1)\\(7 + 2 = 8 + 1)\end{matrix}$$

For the reaction

$$^{238}_{92}U \rightarrow \alpha + X$$

we can substitute $^{4}_{2}He$ for the alpha particle. The equation then becomes

$$^{238}_{92}U \rightarrow {}^{4}_{2}He + X$$

Applying the rule that the sum of the atomic numbers on both sides of the equation must be the same,

on the left side we have 92 and on the right side 2

so that the atomic number of X must be 90.

Likewise, the mass number on the left side is 238 and on the right side 4 so that the mass number of X must be 234. The equation then becomes

$$^{238}_{92}U \rightarrow\ ^{4}_{2}He\ +\ ^{234}_{90}\text{element}$$

From the periodic table we see that the symbol of the element with atomic number 90 is Th (for the element thorium). Thus the equation becomes

$$^{238}_{92}U \rightarrow\ ^{4}_{2}He\ +\ ^{234}_{90}Th$$

Note that the substance produced is an isotope of thorium.

Next consider the reaction

$$^{214}_{82}Pb \rightarrow \beta\ +\ Y$$

Rewriting this equation with $^{0}_{-1}e$ for the beta particle, we have

$$^{214}_{82}Pb \rightarrow\ ^{0}_{-1}e\ +\ Y$$

The atomic number on the left side of the equation is 82 and on the right side it is -1, so that the atomic number of Y is 83. The mass number on the left side of the equation is 214 and on the right side it is 0, so that the mass number of Y is 214. Therefore, the product Y is an element with atomic number 83 and mass number 214. From the periodic table, element 83 is bismuth, Bi.

The equation thus becomes:

$$^{214}_{82}Pb \rightarrow \beta\ +\ ^{214}_{83}Bi$$

Now consider the reaction

$$^{234}_{90}Th \rightarrow \gamma\ +\ ?$$

A gamma ray is an electromagnetic radiation, not a particle. That is, it has no mass (a mass number of 0). It also has no atomic number since it is not an element. We can, therefore, rewrite this equation as

$$^{234}_{90}Th \rightarrow \gamma\ +\ ^{234}_{90}Th^*$$

where we note that the same element, Th, is produced since there is no change in atomic number or mass number. Note however, that there is an asterisk after the product. This is a method of designating that the product has less energy than the reactant (it lost some energy in emitting the gamma ray).

The results of a radioactive change whereby an element loses an alpha particle, a beta particle, or a gamma ray may be summarized as follows:

When an atom emits an α particle, its mass number decreases by 4 and its atomic number decreases by 2 (since the mass of the helium nucleus is 4 and its nuclear charge is +2).

When an atom emits a β particle, its mass number remains the same and its atomic number increases by 1 (since the mass of the electron is zero and its charge = -1).

When an atom emits a γ ray, its mass number and atomic number remain unchanged (since a γ ray has no mass and no charge but does contain energy).

IV. ARTIFICIAL RADIOACTIVITY AND THE TRANSMUTATION OF ELEMENTS

Natural radioactivity is due to the spontaneous changes within the nucleus. We can achieve the same results with non-radioactive substances by means of nuclear bombardment with high speed neutrons, alpha particles, deuterons, or protons. When one of these particles strikes a nucleus, we might consider it as momentarily adding to the nucleus and then causing some further reaction.

| Incoming particle | Parent nucleus | Unstable nucleus | New nucleus | plus ? |

Thus,
when a nucleus gains a neutron, its mass number increases by 1, but its atomic number remains the same since a neutron has a mass of 1 and a charge of zero. A neutron is abbreviated as $_0n$, atomic number zero, mass number 1.

When a nucleus gains an alpha particle, its mass number increases by 4 and its atomic number increases by 2.

When a nucleus gains a deuteron, its mass number increases by 2, and its atomic number increases by 1 since a deuteron has a mass number of 2 and an atomic number of 1. It is abbreviated as $_1H$.

When a nucleus gains a proton, its mass number increases by 1 and its atomic number increases by 1. The proton has a mass number of 1 and an atomic number of 1 so it is abbreviated as $_1H$.

When a nucleus emits an electron, its mass number remains the same but its atomic number increases by 1.

When a nucleus emits a positron, its mass number remains the same but its atomic number decreases by 1. The positron has a mass number of zero and a charge of $+1$ so it is abbreviated as $_1e$.

Example No.24-3.   Complete:

$$_{13}^{27}Al + n \rightarrow \,_{11}^{24}Na + ?$$

Rewriting this equation with $_0^1n$ for the neutron, we have

$$_{13}^{27}Al + \,_0^1n \rightarrow \,_{11}^{24}Na + ?$$

The sum of the mass numbers on the left is 28 (27 + 1) and on the right it is 24, so the missing particle (which we shall call X temporarily) has a mass number of 4, or ^4X. The sum of the atomic numbers on the left is 13 (13 + 0) and on the right it is 11, so X must have an atomic number of 2, or $_2X$. The particle with an atomic number of 2 is Helium, so the correct reaction is

$$_{13}^{27}Al + \,_0^1n \rightarrow \,_{11}^{24}Na + \,_2^4He$$

Example No. 24-4. Complete:

$$\underset{11}{\overset{24}{}}Na \rightarrow \underset{12}{\overset{24}{}}Mg + ?$$

Since the masses are already the same on each side of the equation, the missing particle must have a mass number of zero or ^0X where X represents the symbol for the missing substance. The atomic number on the left is 11 and on the right 12, so the missing substance must have an atomic number of -1, or $_{-1}X$. The substance with an atomic number of -1 is the electron, so the correct equation is

$$\underset{11}{\overset{24}{}}Na \rightarrow \underset{12}{\overset{24}{}}Mg + \underset{-1}{\overset{0}{}}e$$

V. HALF-LIFE

The half-life of a radioactive element is the time required for half of a substance to disintegrate or to change to another substance. For example, the half-life of ^{14}C is 5600 years. This means that if you had 1 gram of carbon^{14}, after 5600 years only ½ gram of the carbon would be left, the rest having been changed into something else. After another 5600 years, only ¼ gram would be left (½ of ½ gram). After yet another 5600 years only ⅛ gram (½ of ¼ gram) would be left.

The time required for radioactive disintegration can be expressed by the formula

$$t = \frac{1}{k}\ln\frac{N_0}{N}$$

where N_0 is the number of atoms (or the number of grams) you have at the beginning, N is the number of atoms (or grams) after time t, and k is a decay constant which is characteristic for each isotope.

Example No. 24-5. $\underset{90}{\overset{234}{}}Th$ has a half-life of 24.1 days. Calculate its decay constant.

Using the formula $t = \frac{1}{k}\ln\frac{N_0}{N}$ where the amount left (N) is ½ of the original amount after one half-life period or 24.1 days, we have

$$24.1 \text{ days} = \frac{1}{k}\ln\frac{N_0}{\frac{1}{2}N_0} = \frac{1}{k}\ln 2 = \frac{1}{k}(0.693) = \frac{0.693}{k}$$

Then, $k = \dfrac{0.693}{24.1 \text{ days}} = 0.0288/\text{day}$

Example No. 24-6. If you had 0.500 g $\underset{90}{\overset{234}{}}Th$ (half-life 24.1 days), how much would be left after 10.0 days?

To calculate the amount left, we need to know the value of k, the decay constant for this particular isotope. We have already calculated it in the previous example, so, using the above formula.

$$10.0 \text{ days} = \frac{1}{0.0288/\text{day}}\ln\frac{0.500 \text{ g}}{N}$$

and $\ln\dfrac{0.500 \text{ g}}{N} = \dfrac{10.0 \text{ days} \times 0.0288/\text{day}}{1} = 0.288$

Taking the antilog of both sides, $\dfrac{0.500 \text{ g}}{N} = 1.33$

and $N = \dfrac{0.500 \text{ g}}{1.33} = 0.376$ g

Example No. 24-7. ^{14}C is used to date wooden objects discovered in archeological excavations. If a wooden object is found to have only 55% of the expected ^{14}C cpm (counts per minute), how old is the object? Half-life of ^{14}C is 5600 years.

First, we have to calculate the value of the decay constant. Using the above formula where N is $\frac{1}{2}N_0$ after 5600 years (one half-life period),

$$5600 \text{ years} = \frac{1}{k}\ln\frac{N_0}{\frac{1}{2}N_0} = \frac{1}{k}\ln 2 = \frac{0.693}{k}$$

$$k = \frac{0.693}{5600 \text{ years}} = 1.24 \times 10^{-4}/\text{year}$$

Then, assuming that the number of cpm is proportional to the number of atoms present (if at N_0 we have 1 cpm, at N we have 55% of 1 cpm or 0.55 cpm) we have

$$t = \frac{1}{1.24 \times 10^{-4}/\text{year}}\ln\frac{1}{0.55} = 4800 \text{ years}$$

Problem Assignment

24-1.   Calculate the binding energy for: (a) $^{2}_{1}H$; (b) $^{18}_{8}O$; (c) $^{239}_{92}U$. . . . . (a. 2.16×10^{11} J/mol)

24-2.   Calculate the energy change in J/mol for the following reactions:

   a. $^{2}_{1}H + ^{3}_{1}H \rightarrow ^{4}_{2}He + ^{1}_{0}n$ . . . . (1.69×10^{12} J/mol)

   b. $^{17}_{8}O + ^{1}_{1}H \rightarrow ^{14}_{7}N + ^{4}_{2}He$ . . . . (1.17×10^{11} J/mol)

   c. $^{222}_{86}Rn \rightarrow ^{218}_{84}Po + ^{4}_{2}He$

   d. $^{40}_{19}K + ^{1}_{0}n \rightarrow ^{40}_{18}Ar + ^{1}_{1}H$

24-3. $^{226}_{88}Ra \rightarrow ^{222}_{86}Rn + ?$ . . . . ($^{4}_{2}He$)

24-4. $^{5}_{2}He \rightarrow ^{4}_{2}He + ?$ . . . . ($^{1}_{0}n$)

24-5. $^{232}_{90}Th \rightarrow ^{232}_{91}Pa + ?$ . . . . ($^{0}_{-1}e$)

24-6. $^{13}_{7}N \rightarrow ^{13}_{6}C + ?$ . . . . ($^{0}_{1}e$)

24-7. $^{27}_{13}Al + ? \rightarrow ^{25}_{12}Mg + ^{4}_{2}He$

24-8. $^{14}_{7}N + ^{4}_{2}He \rightarrow ^{17}_{8}O + ?$

24-9. $^{28}_{13}Al \rightarrow ? + ^{0}_{-1}e$

24-10. $^{7}_{3}Li + ? \rightarrow 2\,^{4}_{2}He$

24-11. $^{30}_{15}P \rightarrow ^{0}_{1}e + ?$

24-12. $^{212}_{84}Po \rightarrow ^{208}_{82}Pb + ?$

24-13. $^{239}_{94}Pu \rightarrow ^{235}_{92}U + ?$

24-14. $^{87}_{36}Kr \rightarrow ? + ^{86}_{36}Kr$

24-15. $^{12}_{6}C + ? \rightarrow ^{13}_{6}C + ^{0}_{1}e$

24-16. $^{37}_{18}Ar + ? \rightarrow ^{37}_{17}Cl$

24-17. $^{239}_{93}Np \rightarrow ? + ^{239}_{94}Pu$

24-18. A radioactive isotope has a half-life of 1.00 year. What fraction of it will be left after (a) 1 year, (b) 2 years, (c) 5 years, (d) 10 years? . . . . (1/2, 1/4, 1/32, 1/1024)

24-19. If 20 mg of a radioactive substance with a half-life of 2.0 hours were present at 8 P.M., how much was present at 8 A.M. of that same day? . . . . (1280 mg)

24-20. $^{24}_{11}Na$ has a half-life of 15.0 hours. What is its decay constant?

24-21. If you had 2.00 g $^{24}_{11}Na$, how much would be left after 6.0 hours? after 25 hours? (see Problem 24-20). . . . . (1.51 g, 0.629 g)

24-22. A piece of charcoal is found to have only 15% of the radioactivity of present day specimen. How old is the charcoal? Half-life of ^{14}C is 5600 years.

24-23. A radioactive sample is found to have 2000 cpm at one time and only 800 cpm 24 hours later. What is the half-life of the sample?

24-24. How long will it take for 7/8 of a sample of $^{210}_{84}Po$ to disintegrate? Its half-life is 138 days.

Chapter 25

COMPLEX COMPOUNDS

A complex compound (frequently called a complex) contains a central ion or atom bonded to several groups called ligands. The number of ligands attached to the central ion or atom is called the coordination number of that ion or atom. That part of a complex compound containing a complex ion is usually enclosed in brackets [ ]. Thus, in the complex $[Cu(NH_3)_4]Cl_2$, the central ion is the Cu(II) ion to which is attached four ligands, the four NH_3 molecules.

NAMING COMPLEX COMPOUNDS

In naming complex compounds, the names of most ligands (except ammonia) end in "o" as indicated in the following table.

Ligand	Name		Ligand	Name
H_2O	aquo		NH_3	ammine
Cl^-	chloro		F^-	fluoro
CN^-	cyano		Br^-	bromo
NO_2^-	nitro		NO_3^-	nitrato
SO_4^{2-}	sulfato		OH^-	hydroxo

To indicate the number of ligands attached to the central ion or atom, we use the following prefixes:

1 mono (understood)	2 di
3 tri	4 tetra
5 penta	6 hexa

A. Rules for Naming Complex Compounds Containing Positively Charged Complex Ions

1. A prefix is used to designate the number of ligands attached to the central ion or atom.

2. Then comes the name of the ligand.

3. If more than one type of ligand is present, the negative one is named first, then the neutral one. If more than one negative ligand is present (or if more than one type of neutral ligand is present) each type is named in alphabetical order with the negative ones being named first.

4. Next comes the name of the central ion or atom with its oxidation state written in Roman numerals (in parentheses) after the name.

5. Finally comes the name of the negative ion attached to the positively charged complex ion.

Let us see how these rules apply in naming the following complex compounds (and ions).

Example No. 25-1.   Name the complex compound: $[Ag(NH_3)_2]Cl$.

We see that there are two ligands, so we use the prefix di. The ligands are NH_3's so we use the name ammine. The central ion is the Ag(I) ion so that the name of this complex is

diamminesilver(I) chloride.

If we consider only the complex ion, $[Ag(NH_3)_2]^+$ its name is

diamminesilver(I) ion.

Example No. 25-2. Name the complex compound: $[Ni(H_2O)_6]SO_4$.

In this compound we see that the ligand H_2O is a neutral molecule and so has no charge. The SO_4 ion has a -2 charge so that the central Ni ion must have a charge of $+2$. The name of this complex is

hexaaquonickel(II) sulfate.

Example No. 25-3. Name the complex: $[Co(H_2O)_4Cl_2]Cl$.

In this complex we must first calculate the oxidation state of the central ion. Since the H_2O ligand is a neutral molecule, there is no charge associated with it. Three chlorides have a total charge of -3, so that the central cobalt ion must have a charge of $+3$.

Therefore, following the rule that the negative ligands are named before the neutral ones, the name is

dichlorotetraaquocobalt(III) chloride.

B. Rules for Naming Complex Compounds Containing Negatively Charged Complex Ions

If a compound contains a negatively charged complex ion, we follow the same rules as before with two exceptions.

1. The positive ions attached to the negatively charged complex ion are named first.

2. The name of the central ion must end in ATE.

Example No. 25-4. Name the complex: $K_3[Fe(CN)_6]$.

To name this complex we must first calculate the oxidation state of the central ion. Six cyanide groups have a total charge of -6; three potassium ions have a total charge of $+3$ so that the iron ion must have an oxidation state of $+3$. The name of this complex is

potassium hexacyanoferrate (III).

If we consider the complex ion only, $[Fe(CN)_6]^{3-}$, its name is

hexacyanoferrate(III) ion.

Example No. 25-5. Name the complex: $Na[Al(H_2O)_2(OH)_4]$.

Following the rules given above and noting that the oxidation state of the central aluminum is $+3$ and naming the negative ligand first, the name of this complex is

sodium tetrahydroxodiaquoaluminate(III).

Example No. 25-6. Name the complex: $K[Co(H_2O)_2(NO_2)_4]$.

In this complex the oxidation state of the central Cobalt ion is $+3$ since the K ion is $+1$, the H_2O's are neutral and the four NO_2's are -4. The name of this complex is

potassium tetranitrodiaquocobaltate(III).

The names of some central ions are derived from their latin names. Thus, for silver we have argentate; for gold, aurate; for copper, cuprate; for lead, plumbate; and for tin, stannate.

C. Naming Neutral Complex Compounds

A neutral complex compound is named by the same system used for naming complexes containing positively charged complex ions.

Example No. 25-7. Name the complex compound: $[Co(NH_3)_3(NO_2)_3]$.

The name of this compound, using the same system we have already discussed, and noting that the oxidation state of the Cobalt is +3, is

trinitrotriamminecobalt(III).

D. Naming Complex Compounds Containing Both Positive and Negatively Charged Complex Ions

Compounds containing both positive and negative complex ions are named by the same system already discussed.

Example No. 25-8. Name the complex compound: $[Ag(NH_3)_2]_4[Fe(CN)_6]$.

Following the rules, and noting that the oxidation state of the central iron ion is +2, the name is

diamminesilver(I) hexacyanoferrate(II)

Example No. 25-9. Name the complex: $[Cu(H_2O)_4][PtBr_4]$.

The name of this complex compound is

tetraaquocopper(II) tetrabromoplatinate(II).

Alternative Prefixes

If the name of a ligand contains a prefix such as "di" in ethylenediamine, then an alternative prefix must be used to designate the number of ligands present in the molecule. Alternative prefixes are:
"bis" for 2
"tris" for 3 and
"tetrakis" for 4

When alternative prefixes are used, the name of the ligand is placed in parentheses.

For example, the name of the complex $[Co(NH_2C_2H_4NH_2)_3]Cl_3$ where the ligand $NH_2C_2H_4NH_2$ is called ethylenediamine is tris(etheylenediamine)cobalt(III) chloride

The formula for this compound may be abbreviated as $[Co(en)_3]Cl_3$ and the name may be abbreviated as tris(en)cobalt(III) chloride.

STRUCTURE OF COMPLEX IONS

Let us consider the element Cd, atomic number 48. Its orbital arrangement of electrons is

$$1s^2 \quad 2s^2 \quad 2p^6 \quad 3s^2 \quad 3p^6 \quad 3d^{10} \quad 4s^2 \quad 4p^6 \quad 4d^{10} \quad 5s^2$$

If we represent the electrons in each of the outer orbitals by arrows with a pair of electrons being designated by a pair of arrows pointing in the opposite direction, we have

Likewise the Cd^{2+} ion will have the orbital arrangement

When this Cd^{2+} ion combines with four NH_3 molecules to form a complex ion $[Cd(NH_3)_4]^{2+}$, each ammonia molecule, whose structure (see page 11-90) is donates its unshared pair of electrons to the central Cd^{2+} ion so 4 NH_3's will fill up four unused orbitals and thus form the following structure

where the dotted lines indicate the orbitals used for bonding. This type of bonding is called sp^3 bonding because one s orbital and three p orbitals are being used.

The electronic configuration of the above complex ion is the same as that of the inert gas xenon, a fact which helps explain the stability of such a complex ion. However, the inert gas structure is not found in all complex ions.

It has been found that sp^3 bonding corresponds to a tetrahedral structure and thus the $[Cd(NH_3)_4]^{2+}$ or tetraamminecadmium(II) ion will consist of four NH_3 molecules arranged tetrahedrally around a central Cd^{2+} ion.

If we consider the complex ion $[Ag(CN)_2]^-$ we see that there are two ligands attached to the central ion. The orbital arrangement of the silver atom is

Ag $1s^2$ $2s^2$ $2p^6$ $3s^2$ $3p^6$ $3d^{10}$ $4s^2$ $4p^6$ $4d^{10}$ $5s^1$

or, showing the electrons in the outer orbitals only,

the arrangement of the electrons in the Ag^+ ion is

For bonding we will use only two orbitals, since there are only two ligands and thus the bonding will be sp, which corresponds to a linear molecule.

The following table shows the various types of bonding and the geometrical structure produced by such bonding.

Orbitals used	Geometrical configuration
sp	linear
sp^2	planar
sp^3	tetrahedral
dsp^2	square planar
dsp^3	pyramidal
d^2sp^3	octahedral
d^4sp^3	dodecahedral

If we consider the complex ion [Ni(CN)₄]²⁻ the electronic configuration of the central Ni atom and ion are:

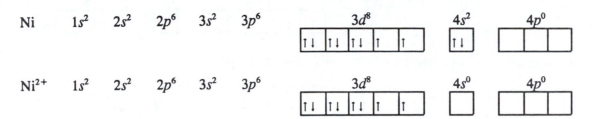

We need four orbitals for bonding the four ligands. We must use only empty orbitals since each ligand supplies two electrons, and this is all any orbital can hold. We should note that in the Ni atom as well as in the Ni ion there are two unpaired electrons. Do these electrons remain unpaired or do they pair up leaving an available 3d orbital? We can answer this question by a study of the magnetic properties of the complex ion. It can be shown that [Ni(CN)₄]²⁻ does not exhibit paramagnetism (magnetic properties) and so has no unpaired electrons. Therefore, the structure of the central Ni ion in this complex ion must be

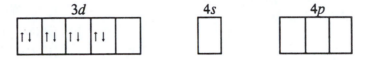

and the next 4 orbitals available for bonding will be a d orbital, an s orbital and 2 p orbitals. This type of bonding, dsp^2, produces a square planar configuration.

If the Ni complex had exhibited paramagnetic properties, as does [Ni(NH₃)₄]²⁺ then there would have been unpaired electrons and the structure of the central Ni ion would have been

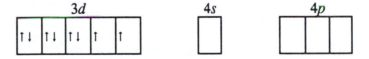

and the bonding would have been sp^3 or tetrahedral

In the case of the hexanitrocobaltate(III) ion, [Co(NO₂)₆]³⁻, where the cobalt atom and the cobalt ion have the following configuration,

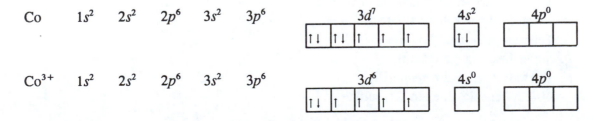

and where the complex does not exhibit paramagnetic properties, the structure of the central ion must be

$$1s^2 \quad 2s^2 \quad 2p^6 \quad 3s^2 \quad 3p^6 \qquad\qquad 3d \qquad\qquad 4s \qquad\qquad 4p$$

The six orbitals available for bonding will be 2 d's, one s and 3 p's or

d^2sp^3 bonding which corresponds to an octahedral structure.

It should be noted that there are other approaches to bonding of coordination complexes. These are discussed in a more advanced text.

Problem Assignment

25-1. Name the following complex compounds:

a. $[Al(H_2O)_6]Br_3$
b. $[Li(NH_3)_4]I$
c. $[Cr(NH_3)_6]Cl_3$
d. $[Pd(NH_3)_2Cl_2]$
e. $K_3[FeF_6]$
f. $K_2[Zn(OH)_4]$
g. $Cu_2[Fe(CN)_6]$
i. $K_2[PtCl_6]$
h. $[Cr(NH_3)_5NO_2]Cl_2$

j. $K_3[Co(NO_2)_6]$
k. $[Pt(NH_3)Cl_5]Cl$
l. $[Cu(NH_3)_4]_3[Fe(CN)_6]_2$
m. $Na[Cu(CN)_2]$
n. $K_2[SnCl_6]$
o. $[Al(H_2O)_6]_2[Zn(OH)_4]_3$
p. $[Ag(NH_3)_2]_2[SnCl_6]$
q. $Zn(en)_2Cl_2$

25-2. Name the following complex compounds or ions:

a. $[Co(H_2O)_6]Cl_3$
b. $[Co(H_2O)_5Cl]Cl_2$
c. $[Co(H_2O)_4Cl_2]Cl$
d. $[Co(H_2O)_3Cl_3]$
e. $[Cu(CN)_4]^{3-}$

f. $[AgCl_2]^-$
g. $[Cd(NH_3)_4]^{2+}$
h. $[Zn(OH)_4]^{2-}$
i. $[Cu(NH_3)_4]^{2+}$
j. $[Co(NH_3)_6]^{3+}$

25-3. Predict the type of structure for the following complex ions:

a. $[Zn(NH_3)_4]^{2+}$
b. $[Fe(CN)_6]^{4-}$ (do not exhibit magnetic properties
c. $[Ni(CN)_4]^{2-}$ (do not exhibit magnetic properties
d. $[Co(NO_2)_6]^{3-}$ (do not exhibit magnetic properties)
e. $[Cu(NH_3)_2]^+$

25-4. Write the formula for each of the following compounds:

a. Potassium tetrahydroxodiaquoaluminate(III)
b. Sodium tetracyanocuprate(I)
c. Zinc hexacyanoferrate(III)
d. Tetrabromodiammineplatinum(VI) bromide
e. Hexaaminnenickel(II) sulfate
f. Tetraaquocadmium(II) nitrate
g. Silver hexacyanoferrate(II)
h. Trichlorotriaquochromium(III)
i. bis(ethylenediamine)zinc(II) sulfate

INDEX